APPALACHIAN TECTONICS

Mountain ranges are the most conspicuous elements of the earth's architecture, and the manner in which the architectural units are arranged or disarranged has become the study of a subdivision of geology known as Tectonics. A hundred years ago James Hall attempted the first scientific synthesis of the steps in the building of the eastern North American mountains, the Appalachians. His initial hypothesis of 1857, expanded and broadened by J. D. Dana during the decade which followed, laid the foundation for our modern geosynclinal theory of mountain building. During the last century modifications and refinements were contributed concerning the roles played by crustal compression, sub-crustal convection currents, batholiths, metamorphism, gravity sliding, and isostasy. In recent years detailed mapping, supplemented by studies of turbidity currents, paleo-magnetism, stable isotopes, and radioactivity have helped to unravel the history of mountain building, but today there are as many questions unanswered as there are those for which there are tentative solutions.

Aspects of Appalachian orogeny was a suitable subject for the symposium of the Royal Society of Canada Annual Meeting in 1966 at Sherbrooke, Quebec—a city within the Appalachian Mountain System. This book assembles the papers of this symposium, dealing with gravity sliding, studies of sedimentation and structure in limited areas, comparisons with the Appalachians of the United States, the bearing of gravity measurements upon our understanding of mountain structure, earthquakes, and a broad, general view of the tectonic pattern of the earth of which this mountain-built belt is but a small part.

Such a comprehensive volume, bringing together a variety of points of view of some of the foremost scholars in the field, indicates the vastness of the subject, the significant progress made thus far, the necessity for new and progressive methods of exploration, and above all the interdependence of all the workers in the field, no matter how seemingly unrelated their specialities are.

The editor of this volume, T. H. CLARK, is Professor, Department of Geological Sciences, McGill University, Montreal.

THE ROYAL SOCIETY OF CANADA

Special Publications

APPALACHIAN
TECTONICS

THE ROYAL SOCIETY OF CANADA
SPECIAL PUBLICATIONS, NO. 10

Edited by Thomas H. Clark

PUBLISHED BY THE UNIVERSITY OF TORONTO PRESS
IN CO-OPERATION WITH
THE ROYAL SOCIETY OF CANADA
1967

PREFACE

THE TITLE of this volume, "Appalachian Tectonics," is disarmingly simple. Only when one attempts to bring together any sort of synthesis between the covers of a small volume do the complications appear. Longitudinally the Canadian Appalachians extend from the international border north and east to Gaspé and on to Newfoundland. Transverse to this trend there lie belt after belt of deformed rocks (e.g. from Rivière du Loup on the south shore of the St. Lawrence through parts of Maine, New Brunswick, and including Nova Scotia) all of which contribute to the Appalachian Complex. One must draw the line somewhere, and we have here chosen to select our subjects from those areas affected by the Taconic and the Acadian orogenies, leaving those areas mountain-built in Late Paleozoic time for another occasion. Even so, it was not possible to attempt a complete coverage of the subject thus limited, though by touching down, so to speak, on half-a-dozen spots some of the myriad problems in tectonics can be brought under scrutiny.

We have thought best to start at the northern extremity of the Appalachians. Early in this volume M. F. Tuke and D. M. Baird, and also L. M. Cumming, discuss some of the problems involving the displacement of klippen in western Newfoundland in which gravity sliding is considered to have played a role. This is referred to later in the paper by M. J. S. Innes and A. Argun-Weston. Although gravity sliding was first proposed well over a century ago it is only in the last thirty years or so that it has become a recognized part of the orogenic cycle and, as with any relatively recent explanation of geological structures, every supporting piece of evidence should be brought forward.

Thence we skip to the Gaspé Peninsula, and the extension of its tectonic elements to the international border. F. Fitz Osborne provides an introduction to this wide and immensely complicated belt, and is followed by four contributions, from north to south, by W. B. Skidmore, Claude Hubert, Pierre St-Julien, and Jacques Béland. The first three of these are direct contributions from members of the Department of Natural Resources of Quebec, and they, together with the fourth, display to good advantage the intensive study of the Appalachian region currently being pursued. And lastly, in this section, is a paper by W. A. Cady whose work in Vermont for the United States Geological Survey brought him frequently into Quebec. His synthesis of the geology north and south of the border is a most welcome addition.

Finally we have three papers that do not have, within the limits of our symposium, specific geographic boundaries. M. J. S. Innes and A. Argun-

Weston bring to bear upon the structure and distribution of rock masses in the Appalachians a valuable mass of data from gravity measurements, corroborating some geological conclusions, challenging others, and pointing the way to future field research. W. E. T. Smith presents a comprehensive review of the distribution and significance of historic earthquakes, and concludes that no relationship has been found to exist between such earthquakes and faults. Lastly J. T. Wilson takes a broad view of the eastern margin of North America and relates deformation of the Atlantic Ocean to Appalachian Tectonics.

THOMAS H. CLARK

CONTENTS

CONTRIBUTORS

A. ARGUN-WESTON, *Dominion Observatory, Ottawa, Ontario*

D. M. BAIRD, *Director, Museum of Science and Technology, Ottawa, Ontario*

JACQUES BÉLAND, *Département de géologie, Université de Montréal, Montréal, Québec*

WALLACE M. CADY, *United States Geological Survey, Denver, Colorado*

THOMAS H. CLARK, F.R.S.C., *Department of Geological Sciences, McGill University, Montréal, Québec*

L. M. CUMMING, *Palaeontology Section, Geological Survey of Canada, Ottawa, Ontario*

CLAUDE HUBERT, *Département des richesses naturelles, Québec, Québec*

M. J. S. INNES, *Dominion Observatory, Ottawa, Ontario*

F. FITZ OSBORNE, F.R.S.C., *Département de géologie, Université Laval, Québec, Québec*

PIERRE ST-JULIEN, *Département des richesses naturelles, Québec, Québec*

W. B. SKIDMORE, *Département des richesses naturelles, Québec, Québec*

W. E. T. SMITH, *Dominion Observatory, Ottawa, Ontario*

M. F. TUKE, *Department of Geology, Ottawa University, Ottawa, Ontario*

J. T. WILSON, F.R.S.C., *Institute of Earth Sciences, University of Toronto, Toronto, Ontario*

APPALACHIAN TECTONICS

KLIPPEN IN NORTHERN NEWFOUNDLAND

M. F. Tuke and D. M. Baird

ABSTRACT

Two contemporary rock sequences outcrop in northern Newfoundland. An allochthonous sequence of slate, greywacke, and volcanic and ultrabasic rock occurs in three slices and lies on shales which rest conformably on a shelf sequence widespread to the west. The trends of slickensides and the dip of axial planes indicate movement toward the northwest. The allochthonous rocks are thought to have moved by gravity sliding from a source about 60 km to the southeast.

CAMBRIAN AND ORDOVICIAN rocks in western Newfoundland belong to two contrasting facies; a shelf facies of quartz sandstone and carbonate; and a eugeosynclinal facies of slate, greywacke, and volcanic and ultrabasic rocks. Early workers placed all the beds in one sequence but this became unsatisfactory when fossil evidence indicated that some of the beds of very different lithologies were contemporary. In 1963 Rodgers and Neale proposed that the beds belonged to two sequences one of which, originally deposited somewhere to the east, had moved by gravity sliding into its present position.

In 1937 Cooper mapped Hare Bay in the northeast part of the Great Northern Peninsula and found Ordovician carbonates, shales, greywackes, volcanic beds, and one major low-angle dislocation with a displacement of about 15 km. On the basis of this report Rodgers and Neale suggested that the beds in this northernmost part of the Great Northern Peninsula might be similar to those on the west coast and thus also belong to two sequences, one autochthonous and the other allochthonous.

During the summers of 1963 and 1964 Tuke mapped the area north of Hare Bay to investigate this possibility and presented his results in a Ph.D. thesis under the direction of Baird in 1966. This paper embodies the major conclusions reached but the authors hope to publish a more detailed account of the area later.

The presence of two sequences in the area is confirmed by the existence of coeval beds of very different lithological associations. The first sequence consists of four formations, the Bradore, St. George, Table Head, and Goose Tickle formations. The Bradore Formation, Lower Cambrian quartz sandstone, is found on islands off the east coast (Fig. 1). Limestone and dolomite of the St. George Formation, Lower Ordovician, form the western part of the area. This formation is conformably overlain by limestone of the Middle Ordovician Table Head Formation and this in turn is disconformably overlain by shale and greywacke belonging to the Middle Ordovician Goose

Tickle Formation. The last three formations have sedimentary contacts throughout and must therefore belong to one sequence. This sequence shows an over-all deepening of the depositional surface; the St. George Formation, at the bottom, contains stromatolites and must therefore have been deposited in the photic zone, whereas the Goose Tickle Formation, at the top, consists of basinal sediments. This is the autochthonous sequence which extends widely over western Newfoundland and adjacent Labrador.

The second sequence contains three formations, the Goose Cove, Maiden Point, and Northwest Arm formations. Volcanic beds of the Goose Cove Formation, now metamorphosed by the intrusion of ultrabasic bodies, rest on a major thrust plane and are well exposed on the north side of Hare Bay. The Maiden Point Formation consists of coarse greywacke, slate, and volcanic rocks and is also underlain by a major thrust. Both these formations are thought to be Lower Ordovician or older because they are intruded by peridotites; in other parts of Newfoundland an Ordovician date best suits the occurrence of ultrabasic masses and the nearest radiometrically dated peridotite in the northern Appalachians, Mount Albert, was intruded during the Lower Ordovician (MacGregor 1963). The Northwest Arm Formation, the remaining member of the second sequence, outcrops in fault slices in Pistolet Bay and Hare Bay. It has yielded Lower Ordovician graptolites and is thus contemporaneous with the St. George Formation of the autochthonous sequence. It consists of black and green shale with greywacke beds, now mostly in the form of boudins. The distribution of the formations is shown on Figure 1.

The second sequence is considered to be allochthonous because the Northwest Arm Formation is coeval with part of the autochthonous sequence but consists of a very different lithology. Further, the Goose Cove and Maiden Point formations are underlain by major low angle faults and contain volcanic beds whereas no trace of volcanic material is found in the adjacent autochthonous sequence.

The structural geology is summarised on Figure 2. Most significant are two major low-angle faults which have minimum displacements of 4 and 20 kms. Much of the present outcrop pattern is controlled by normal faults which formed after the arrival of the allochthon.

None of the allochthonous formations are in sedimentary contact with each other; they are thus thought to form three separate klippen.

If some of the beds are allochthonous then it is important to find their origin. Slickenside trends were measured in a number of locations and have preferred trends to the northwest (Fig. 2). Axial planes of the folds dip between southeast and east-southeast. Where measured together, fold axes are at right angles to the trend of the slickensides thus suggesting that they are genetically related. The southeast-northwest trend of the slickensides and the southeast dip of the axial planes suggest that the movement of the beds was to the northwest and thus that the allochthon has moved from the southeast.

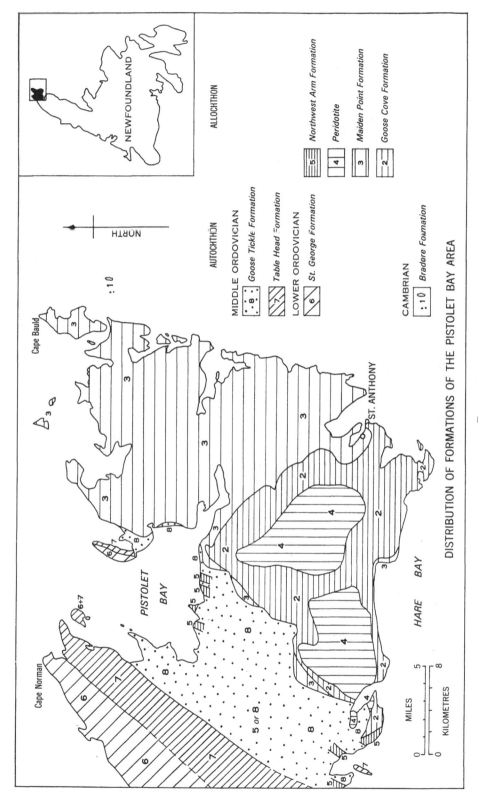

FIGURE 1

DISTRIBUTION OF FORMATIONS OF THE PISTOLET BAY AREA

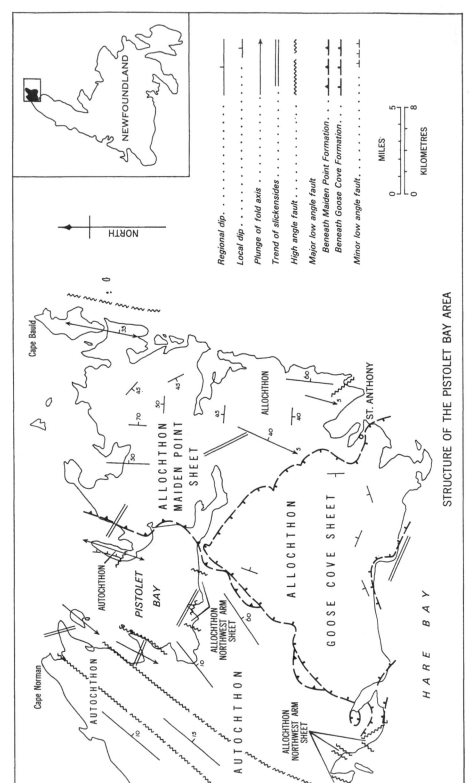

NEWFOUNDLAND

NORTH

Regional dip. . . . .
Local dip. . . . .
Plunge of fold axis . . . .
Trend of slickensides . . . .
High angle fault. . . . .
Major low angle fault
Beneath Maiden Point Formation. . . .
Beneath Goose Cove Formation. . . . .
Minor low angle fault. . . . .

MILES

KILOMETRES

Cape Bauld

33

45

45

50

70

50

ALLOCHTHON
MAIDEN POINT
SHEET

45

40

40

5

ALLOCHTHON

60

45

40

5

ST. ANTHONY

Cape Norman

AUTOCHTHON

AUTOCHTHON

PISTOLET
BAY

ALLOCHTHON
NORTHWEST ARM
SHEET

AUTOCHTHON

60

10

ALLOCHTHON

GOOSE COVE SHEET

ALLOCHTHON
NORTHWEST ARM
SHEET

H A R E   B A Y

10

15

10

STRUCTURE OF THE PISTOLET BAY AREA

FIGURE 2

In central Newfoundland, ultrabasic bodies outcrop in two belts. If the beds around Hare Bay are allochthonous then it is probable that the ultrabasic bodies now found in Hare Bay once formed part of one of the lines of ultrabasic bodies to the south. This places the source area for the allochthonous rocks about 60 km southeast of their present location (Fig. 3).

Some idea of the lithologies that were once present in the source area can be obtained by examining the beds to the south which strike towards the proposed source area. On the east side of the eastern ultrabasic belt is the Fleur de Lys group. This consists of biotite-muscovite-quartz-plagioclase schists and gneisses with minor amounts of amphibolite and marble (Neale and Nash 1963, p. 8). Before metamorphism these rocks may have been shales and greywackes with minor volcanic rocks and limestone. This assemblage is very similar to the lithology of the Maiden Point Formation. Similar rocks but without marble or amphibolite outcrop on the Groais Island, just to the southwest of the proposed source area (Baird 1966, p. 249).

If the allochthonous beds had been pushed into their present position, one would expect them to be more tightly folded than they are. Gravity sliding is preferred to horizontal compression as a mechanism of movement because it does not involve the great compressive stresses to which the rocks would be subjected if the beds had been pushed from behind. Gravity anomalies in the eastern part of the area (oral communication, D. Weaver, Dominion Observatory of Canada) indicate that the basement rises to the east thus suggesting that the necessary slope for gravity sliding existed. Further, the autochthonous sequence shows that the basin of deposition became progressively deeper during the Ordovician thus allowing the allochthon to slide onto it. The Northwest Arm Formation is for the most part a tectonic breccia with boudins of greywacke in a matrix of black and green shale. These beds appear to have been deformed before consolidation. Beds of very similar lithology occur in the classical area of gravity sliding in the Apennines. A breccia composed of fragments of black and green shale and limestone outcrops at the top of the Goose Tickle Formation. These fragments are thought to have been derived from the advancing gravity slide of the Northwest Arm Formation. If this is so the movement of the allochthon is dated as Middle Ordovician. The arrival of the allochthon must be later than lower Middle Ordovician because the beds on which the allochthon rests are of that age. Further evidence for a Middle Ordovician age of movement is found on the west coast where Rodgers (1965) has reported that the Long Point Formation (Middle Ordovician) rests unconformably on rocks of the allochthonous Humber Arm beds. This means that the west coast klippe moved during the Middle Ordovician. It seems likely that the northern allochthon arrived at much the same time.

It is thus thought that the Goose Cove, Maiden Point, and Northwest Arm formations are allochthonous, and that at least the last two were originally deposited 60 km to the southeast but during the Middle Ordovician moved by gravity sliding to their present positions.

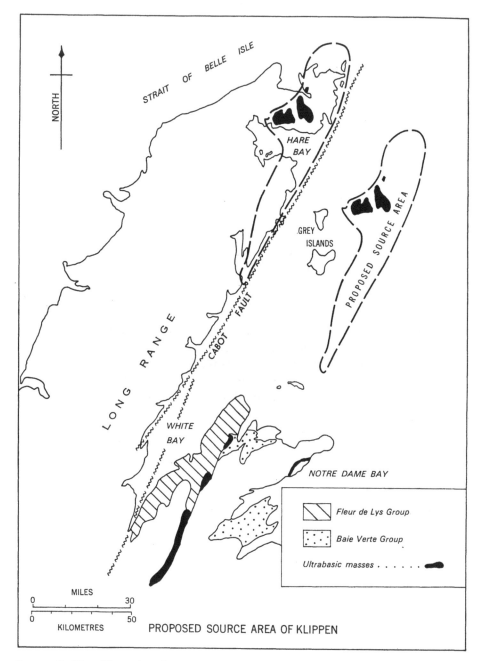

FIGURE 3. The klippe has been moved backward along the direction given by the slickensides to a position where the White Hills ultrabasic masses are aligned with those on the northeast coast.

The great fault in western Newfoundland, recently incorporated by Wilson (1962) in his Cabot fault, separates the proposed source area of the klippen and their present sites. Movement on the fault has been established as pre-, intra-, and post-Carboniferous (Baird 1966, p. 256). This means that the present position of the source area may be to the north-northeast or south-southwest of the position indicated on Figure 3, depending on the sense of movement of the fault.

Current directions in the Goose Tickle Formation are to the southwest, which suggests that the Long Range had not been uplifted in its present form by Middle Ordovician time. The following sequence of events thus seems likely:

1. Accumulation of Cambrian and Ordovician carbonates and quartz sandstones on the shelf to the west and of shales, greywackes, and volcanic rocks to the east and southeast.

2. Intrusion during the Lower Ordovician of ultrabasic masses in the southeast.

3. Deepening of the shelf area and initiation of gravity sliding toward the northwest of large masses of recently deposited rocks and accompanying accumulation of slide-front breccias during the Middle Ordovician.

4. Major movement along the Cabot fault and uplift of the Long Range. Formation of normal faults.

5. Deposition during later periods (e.g. Carboniferous) followed by erosion to establish present outcrop pattern.

REFERENCES

BAIRD, D. M. (1966). Carboniferous rocks of the Conche-Groais Island area. Can. J. Earth Sci., *3*: 247.

COOPER, J. R. (1937). Geology and mineral deposits of the Hare Bay area. Newfoundland Dept. of Nat. Res. Geol. Sec. Bull., *9*: 1–36.

MACGREGGOR, I. D. (1963). Geology, petrology, and geochemistry of the Mount Albert and associated ultramafic bodies of central Gaspé. Can. Min. J., *84* (7): 154.

NEALE, E. R. W. and NASH, W. A. (1963). Sandy Lake (east half) Newfoundland. Geol. Surv. Can., Paper 62–28.

RODGERS, JOHN (1965). Long Point and Clam Bank formations, western Newfoundland. Geol. Assoc. Can., *16*: 83.

RODGERS, JOHN and NEALE, E. R. W. (1963). Possible "Taconic" klippen in western Newfoundland. Amer. J. Sci., *261*: 713.

WILSON, J. T. (1962). Cabot Fault, an Appalachian equivalent of the San Andreas and Great Glen faults and some implications for continental displacement. Nature, *195*: 135.

PLATFORM AND KLIPPE TECTONICS OF WESTERN NEWFOUNDLAND: A REVIEW

L. M. Cumming

ABSTRACT

Parts of two contrasting tectono-stratigraphic regions dominate the geology of western Newfoundland. The Anticosti-Strait of Belle Isle Platform is characterized by relatively undeformed Lower Palaeozoic carbonate rocks, which rest upon Late Precambrian crystalline basement rocks. Superimposed on the margin of the platform rocks is a structurally more complex sequence of klippe rocks, which represent an Appalachian eugeosynclinal environment derived from central Newfoundland. Flat-lying Carboniferous sedimentary rocks overlap onto the margin of the platform on Port au Port Peninsula. To the south, Carboniferous sedimentary rocks are folded parallel to the margins of a northeast-trending trough.

THE RECENT INTEREST in the tectonics of western Newfoundland began ten years ago, following publication of geological maps of Newfoundland by D. M. Baird (1954) and L. J. Weeks (1955). The klippe hypothesis of Rodgers and Neale (1963) has more recently focused attention on this area, which has become something of a classic region to the present generation of North American geologists, through the work of Schuchert and Dunbar (1934, p. 1) who rightly credited James Richardson with providing the foundation for stratigraphic studies in western Newfoundland. From 1860 to 1862, Richardson, as a field geologist for the Geological Survey of Canada, carried out a reconnaissance mapping programme of western Newfoundland.

The area between the Strait of Belle Island, Port au Port Peninsula, and Anticosti Island represents a single structural unit, an Appalachian Platform, where Precambrian rocks are basement. Superimposed on the rim or margin of this platform are klippe rocks, which have been moved from the Appalachian eugeosynclinal zone to the southeast. These tectonically emplaced klippe rocks appear to occupy only a part of the southeastern margin of this platform. Since the western or leading edge of the klippe probably does not extend beyond the near-coastal line indicated by Rodgers and Neale (1963, fig. 1, p. 714), the offshore region of the Gulf of St. Lawrence may offer a potential area for exploration of gas and oil. This offshore region lies within a miogeosynclinal megatectonic unit, as defined by Kuendig (1960, fig. 1).

Geosynclinal rocks of the west coast of Newfoundland are the volcanic and sedimentary units shown as Oh on Baird's map (1954), as well as ultramafic intrusive rocks (shown on Baird's map as Bg, gabbro; Us, serpentinized ultrabasics). The former is the Humber Arm Group and the latter the intrusive Bay of Islands igneous complex (Smith 1958). It is the Rodgers and Neale hypothesis that the root zone for these rocks occurs 40 to 60 miles to the east. There, a thin and discontinuous line of ultramafic rocks extends southwest from Baie Verte. The fiord country of western Newfoundland (Bonne Bay and Bay of Islands) and the coastal area south to the Lewis Hills represent the central and thickest part of the klippe lens. The lens thins towards Daniels Harbour, to the north, and towards Port au Port Peninsula, to the south.

The klippe rocks of the upper Humber Arm terrane at Black Point, 1.5 miles north of Port au Port, are contorted shales and argillites. The axial planes of folds in these rocks are parallel to the contact with the platform carbonates, which lie to the east. Below these rocks in the Port au Port area, and elsewhere in the Humber Arm Group, are good exposures of wildflysch (or a rubble zone). These consist of brecciated blocks of sandstone or greywacke, which have been incorporated into a matrix of Humber Arm shales. A rubble zone is typically developed near the base of the klippe.

Northern Coastal Lowlands

The platform sediments and structures provide a striking contrast with those of the klippe. Flat-lying beds occur on the Labrador coast, on the south side of the Strait of Belle Isle, and on St. John Island in St. John Bay (Fig. 1). As a basis for discussing structures, the stratigraphy of the northwest coastal lowlands of Newfoundland will be outlined. Radiometric age dates of 960 m.y. from the central core of the Long Range, east of St. John Bay, show that the basement rocks are of Grenville age (Poole, Kelley, and Neale 1964, fig. 4).

The Lower Cambrian Labrador Group has a widespread distribution in northwestern Newfoundland (Fig. 1). A complete section of the Labrador Group, well over 1,000 feet thick, may be obtained from exposures at Bradore, Forteau, and Hawke bays. A succession, typical of the sandy facies of this group, consists of a lower red orthoquartzitic sandstone, a middle grey limestone and shale unit with archaeocyathid reef (the Forteau Formation), and an upper white orthoquartzitic sandstone (the Hawke Bay Formation). This succession is an excellent example of a complete cycle of transgressive and regressive sedimentation in a platform environment.

The section continues with the Cambro-Ordovician St. George dolomite and limestone. This unit is 2,000–3,000 feet thick. Along the Strait of Belle Isle it is largely dolomite. Structurally it is flat-lying or with gentle dips and has a pronounced set of vertical joints. The uppermost part of the St. George is a thin-bedded buff dolomite. This lithology is consistent for 250 miles along the west coast of Newfoundland, indicating the stability of the depositional environment. The contact of this thin-bedded buff dolomite with the

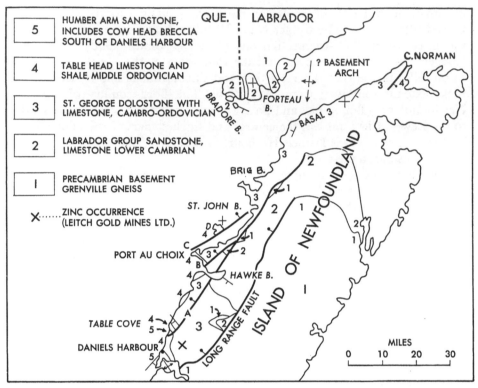

FIGURE 1. Sketch map of regional structure of the coastal lowlands, northwestern Newfoundland.

overlying Table Head Formation is a disconformity. At the Aguathuna Quarry, Port au Port Peninsula, the Table Head black limestone fills a channel 30 feet deep in the buff dolomite of the St. George Formation. The zinc occurrence 7 miles northeast of Daniels Harbour (Fig. 1) is in dolomite breccias, between two and three hundred feet stratigraphically below the general level of the St. George-Table Head contact. This is a Mississippi Valley type of deposit, which has similarities with platform deposits in Missouri and at Pine Point, N.W.T.

The Table Head Formation consists of grey limestone and dolomite and dark grey shale and is of Middle Ordovician age (Whittington 1965). The formation is at least 1,100 feet thick at the Table Cove type section and represents a peripheral miogeosynclinal facies (Ross 1964, p. 1535). Because of the disruptive effect of the klippe, the uppermost beds of the Table Head Formation are not clearly represented in exposures along the west coast of Newfoundland. Additional beds, which perhaps represent the uppermost part of the Table Head Formation, may occur in the offshore region.

The platform sediments are faulted, and the degree of faulting appears to be much greater towards the outer, or seaward, margin of the platform.

On Anticosti Island, faults are so rare that a displacement of even a foot or so has merited special comment (Twenhofel 1928). There is, however, one well-documented, sizeable fault on Anticosti Island. This is a normal fault 8 miles west of East Point, and has a stratigraphic displacement of 40–50 feet, with the east side downthrown. This fault was described by James Richardson (1856). On the west coast of Newfoundland, some of the faults in the platform sediments are of the reverse type, which gives a shingling effect and locally produces repetitions of stratigraphic units. The Cape Cormorant sea-cliffs, 700 feet high, on Port au Port Peninsula show reverse faults. Also, the parallel faults mapped by Baird (1960) in the Parsons Pond-Western Brook Pond region appear to be reverse faults.

Normal faults are also present in the platform sediments. A pattern of four subparallel major normal faults is shown on Figure 1. These faults are the Long Range fault and faults A, B, and C. Each of these four faults shows uplift on the east side, emphasizing the uplift of the Precambrian core of the Long Range. The Table Head-St. George contact is stepped to the south in successive fault blocks.

The Table Head-St. George contact is moved by Fault C south to Port au Choix, from a position along the north side of St. John Island. Fault B moves the contact farther south to a position 10 miles north of Daniels Harbour. Fault A moves the contact about 4,000 feet farther south, so that the offset at the coast is approximately that of the width of the coastal out-crop belt of the Table Head Formation. There are minor normal faults in this area as well. One example, too small to be shown on Figure 1, trends northeast across the neck of the Port au Choix Peninsula. This fault, and others in the same area, is marked by an increase in dolomite content towards the fault plane. A shallow basement arch is indicated on Figure 1 in the Strait of Belle Isle region. Evidence for this basement structure is:

1. A normal fault near Cape Norman east of the arch, with the east side downthrown. Faults on the west side of the arch have the west sides downthrown.

2. The upper (Lower Ordovician) part of the St. George Formation is exposed near Cape Norman and Brig Bay, while the lower (Middle and Upper Cambrian) part is exposed along the narrows of the Strait of Belle Isle. Whittington and Kindle (1966) recently established the Cambrian age of these lower beds.

3. The distribution of Lower Cambrian strata in Labrador reflects the position of the arch. These beds occur at Forteau Bay and Battle Harbour, but are missing in the intervening area.

PORT AU PORT REGION

In the Port au Port Peninsula region, platform carbonates of the St. George and Table Head formations dip 5 to 20 degrees to the north (Fig. 2). That is, they dip under, or are mantled by, the folded red and green

shales, sandstones, and argillites of the Humber Arm Group, which lie to
the north. These beds, which are well exposed along the shores of Middle
Point in Port au Port Bay, are correlated with the "Upper Humber Arm
terrane" of Lilly (1964). Sandstones at Black Point, East Bay (the eastern
F2 locality on Fig. 2) are correlated with the "Western Sandstone Formation" of the Bay of Islands region (Lilly 1963).

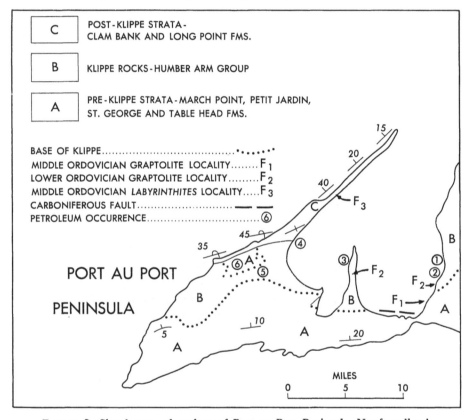

FIGURE 2. Sketch map of geology of Port au Port Peninsula, Newfoundland.

The succession (A, B, C) shown on Figure 2 is a tectonic succession,
but not a succession of ages. Klippe rocks are sandwiched between platform
carbonates, and there is an age inversion as Lower Ordovician rocks rest on
(or mantle) Middle Ordovician rocks. The Table Head Formation (at F1)
contains *Climacograptus* of Middle Ordovician (Whiterock) age. The
Humber Arm shales (at F2) contain *Adelograptus* and *Bryograptus* of
Lower Ordovician (Hunneberg) age. The Long Point Formation (at F3)
contains *Labyrinthites* (Bolton 1965) of Middle Ordovician (Wilderness)
age.

This new faunal evidence about the timing of the klippe emplacement,
plus confirmation that the upper contact of the klippe with the Long Point

Formation is a depositional contact (Rodgers 1965, p. 84), shows that the klippe emplacement was entirely a Middle Ordovician event, and that the platform margin subsided rapidly in Middle Ordovician time. This subsidence, coupled with uplift in central Newfoundland, would produce the necessary energy gradient for gravity sliding to take place, as postulated by Rodgers and Neale (1963).

One of the interesting aspects of this age inversion on Port au Port Peninsula is that the klippe rocks have acted as a cap rock to trap petroleum generated from a source in the underlying platform carbonates. Six types of petroleum occurrences in the Port au Port area appear to be related to a capping effect of the impervious beds near the base of the klippe (Cumming 1965).

Figure 3 is a cross-section through Anticosti Island and Port au Port Peninsula. The line of section has been chosen to cut across the southern part of the klippe. The distance across the Gulf of St. Lawrence is 125 miles, and the possibility of Late Palaeozoic tectonics in this offshore region has not been taken into consideration in constructing this section. Lower Cambrian sediments pinch out to the west. The Romaine and upper St. George carbonate unit (O1 on Fig. 3) extends across the platform, as does the disconformity above it. The klippe rocks (part of which represent a shale facies of the same age as carbonates within the upper part of the St. George Formation) are shown as if they acted as a thick stratigraphic unit, after once being emplaced. Their basal contact has the form of a low-angle thrust; their upper contact is a depositional contact. The Becscie Formation and

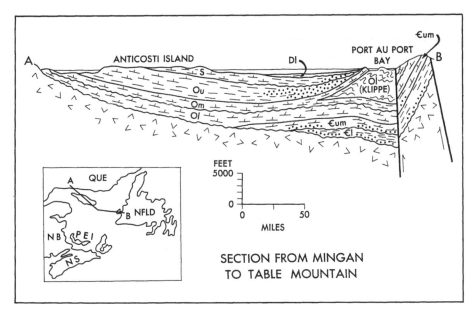

FIGURE 3. Cross-section through Anticosti Island and northern part of Port au Port Peninsula.

younger Silurian carbonates of Anticosti Island are not represented on Port au Port Peninsula. However, a red-bed facies of Silurian age may be represented beneath the fossiliferous upper part of the Lower Devonian Clam Bank Formation (D1 on Fig. 3). This cross-section points out the probable limited lateral extent of the klippe structure as compared with the widespread platform structure.

Also in Carboniferous time, the platform influenced sedimentation and deformation, Mississippian conglomerates form the top of the sea-cliffs and rest unconformably on Ordovician beds at the west end of Port au Port Peninsula (Riley 1962). To the south (south of the head of St. Georges Bay), Pre-Carboniferous (Palaeozoic) rocks form the basement (Howie and Cumming 1963). These younger basement rocks are composed of Devonian or earlier gneisses, schists, and granitic rocks, with age dates of 400 and 415 m.y. (Gillis 1965). In this area of southwestern Newfoundland, Carboniferous rocks are represented by a thick succession of clastic sediments (Baird and Cote 1964). These Carboniferous sediments, in contrast to the flat-lying Carboniferous beds at Port au Port Peninsula and around the head of St. Georges Bay, are folded parallel to the sides of a northeast-trending trough (Neale, Béland, Potter, and Poole 1961).

In summary, this review of the tectonics of the west coast of Newfoundland has stressed the importance of a stable platform region. Klippe rocks appear to be limited to a narrow margin of this broad structural unit. This platform was initiated as an early Palaeozoic structure and maintained its relative stability throughout the Palaeozoic Era.

REFERENCES

BAIRD, D. M. (1954). Geological map of Newfoundland. Geol. Surv. Newfoundland.
——— (1960). Sandy Lake, west half, Newfoundland. Geol. Surv. Can., Map 47–1959.
BAIRD, D. M. and COTE, P. R. (1964). Lower Carboniferous sedimentary formations in southwestern Newfoundland and their relations to similar strata in western Cape Breton Island. Bull. Can. Inst. Min. and Met., 57: 509–519.
BOLTON, T. E. (1965). Ordovician and Silurian tabulate corals: *Labyrinthites, Arcturia, Troedssonites, Multisolenia,* and *Boreaster.* Geol. Surv. Can., Bull. 134: 22–23.
CUMMING, L. M. (1965). Biostratigraphic studies, Port au Port area, in Report of activities: field, 1964, *compiled by* S. E. Jenness, Geol. Surv. Can., Paper 65–1: 104.
GILLIS, J. W. (1965). Port aux Basques, 110, Map-Area *in* Report of activities, field, 1964 *compiled by* S. E. Jenness, Geol. Surv. Can., Paper 65–1: 133–135.
HOWIE, R. D. and CUMMING, L. M. (1963). Basement features of the Canadian Appalachians, Geol. Surv. Can., Bull. 89.
KUENDIG, E. (1960). Eugeosynclines as potential oil habitats. Fifth World Petroleum Congress, New York, 1959, Proceedings, Section I, Paper 25: 461–479.
LILLY, H. D. (1963). Geology of Hughes Brook-Goose Arm area, west Newfoundland, Geol. Rep. 2, Memorial University of Newfoundland.
——— (1964). Possible "Taconic" klippen in western Newfoundland. Am. Jour. Sci., 262, Discussions 1130–1135.
NEALE, E. R. W., BÉLAND, J., POTTER, R. R., and POOLE, W. H. (1961). A preliminary tectonic map of the Canadian Appalachian region based on age of folding. Bull. Can. Inst. Min. and Met., 54: 687–94.
POOLE, W. H., KELLEY, D. G. and NEALE, E. R. W. (1964). Age and correlation problems in the Appalachian region of Canada. *In* Geochronology in Canada, *edited*

by F. Fitz Osborne, Roy. Soc. Can. Spec. Publ. *8*: 61–84. Toronto. University of Toronto Press.

RICHARDSON, JAMES (1856). Geological map of Anticosti Island, scale 1 inch to 1 nautical mile. Unpublished ms., Geol. Surv. Can., Map 3478.

RILEY, G. C. (1962). Stephenville map-area, Newfoundland. Geol. Surv. Can., Mem. *323*: 72 p.

RODGERS, JOHN and NEALE, E. R. W. (1963). Possible "Taconic" klippen in western Newfoundland. Am. Jour. Sci., *261*: 713–730.

RODGERS, JOHN (1965). Long Point and Clam Bank formations, western Newfoundland. Geol. Assoc. Can., *16*: 83–94.

ROSS, R. J. JR. (1964). Relations of Middle Ordovician time and rock units in basin ranges, western United States. Bull. Am. Assoc. Pet. Geol., *48*(9): 1526–1554.

SCHUCHERT, C., and DUNBAR, C. O. (1934). Stratigraphy of western Newfoundland. Geol. Soc. Am., Mem. *1*: 1–123.

SMITH, C. H. (1958). Bay of Islands igneous complex, western Newfoundland. Geol. Surv. Can., Mem. *290*: 132 p.

TWENHOFEL, W. H. (1928). Geology of Anticosti Island. Geol. Surv. Can., Mem. *154*: 13.

WHITTINGTON, H. B. (1965). Trilobites of the Ordovician Table Head Formation, western Newfoundland. Bull. Mus. Comp. Zool., *132* (4): 277–441.

WHITTINGTON and KINDLE, C. H. (1966). Middle Cambrian strata at the Strait of Belle Isle, Newfoundland, Canada. Abs., Geol. Soc. Am., Program Annual Meeting Philadelphia, Pa., p. 46.

WEEKS, L. J. (1955). Geological map of the island of Newfoundland. Geol. Surv. Can., Map 1043 A.

APPALACHIAN REGION OF QUÉBEC: SOME ASPECTS OF ITS DIVISIONS AND GEOLOGY

F. Fitz Osborne, F.R.S.C.

ABSTRACT

A review of the nomenclature of the longitudinal divisions of the Appalachian region of Quebec shows such glaring inconsistencies of usage that a new series of names is justifiable. The informal transverse divisions are, however, more reasonably serviceable for the geologist.

The present interpretation of the geology is essentially that postulated by Logan in 1843 but abandoned by workers from 1890 until 1950.

THE APPALACHIAN REGION of Quebec is the northwestern part of a segment of the region that extends through Newfoundland to Alabama. Within this great region there is substantial diversity so that the geology of one region differs from that of another. However, "problems" and controversies are perhaps characteristic of the geology of the Appalachian region as a whole, and Quebec has its share of both. The "Taconic question," as it overflowed into Quebec, and the "Quebec Group" are two examples, but Quebec can also provide data that serve to clarify some aspects of Appalachian geology.

LIMITS AND DIVISIONS

The limits of the Appalachian region of Quebec are mostly political or are covered by water, either of the Atlantic ocean or of the estuary of the St. Lawrence River. It is worth comment that a political boundary with the elements of provincialism or nationalism may, through adherence to a local nomenclature or interpretation, foster "problems" that are almost as difficult to deal with as true geological problems.

The western limit of the Appalachian region in Quebec is on land for about 125 miles: from Lake Champlain to Quebec City; but throughout most of this distance the cover of unconsolidated material is so continuous that relatively few direct observations are possible (Clark 1951). However, evidence is accumulating that the western limit is complex and that "Logan's line" perhaps should be replaced, at least in thought, by Logan's belt or zone, and many names that have been given to the great "fault" should perhaps be allowed to lapse.

The Appalachian region of Quebec is 480 miles long. A segment south of the Etchemin River has been long recognized as a unit and is referred to as the Eastern Townships or l'Estrie, and a segment east of Matapédia

River was known even in the sixteenth century as Gaspé. The segment between these has no generally recognized name: "South Shore," which has been suggested, is a colourless appelation. The division is here referred to as the Etchemin-Matapédia segment. Table I shows the lengths of the segments, areas, and trends of the principal folds at the ends of each segment.

TABLE I

SEGMENTS OF THE APPALACHIAN REGION IN QUEBEC

Segment	Length (*miles*)	Area (*sq. miles*)	Trend	
			On S	On N or E
Eastern Townships	128	10,070	N10°E	N50°E
Etchemin-Matapédia	204	9,350	N50°E	N70°E
Gaspé	148	10,240	N20°E	N70°N

It is to be noted that these units have substantial areas: each is about as large as Vermont or New Hampshire. Such an area can display a diversity of geology but, peculiarly, the three rather randomly bounded units have relatively consistent geology. In part this may be a result of the history of investigation, but in a considerable measure it is real.

The longitudinal divisions of the Appalachian region in Quebec have not been agreed upon. "Monts Notre-Dame" was in use before the beginning of the seventeenth century (Trudel 1961), and was applied to a part of the Gaspé peninsula. The name Shickshock has been applied more consistently to a separate range or a part of the Notre-Dame range in Gaspé. However, the usage has not been consistent; for example, Notre-Dame Mountains extend through Gaspé to somewhat south of Thetford Mines on the map of the physiographic divisions in the *Atlas of Canada* (1957). On the same map the Sutton range is also called Green Mountains and is shown as distinct from the Notre-Dame range.

Jules Marcou (1855) presented a classification of the mountain ranges of part of North America ostensibly after the system used by Elie de Beaumont in France. Indeed he states (Marcou 1855, p. 329) that the prolongations of two of the European mountain ranges coincide "most perfectly" with two ranges in North America. The "Laurentine mountains" trend principally east-west and are developed in the Precambrian shield but include terrane between Moosehead Lake and the Saint John River in Maine as well as mountains north of Bathurst in New Brunswick. Marcou (1855, p. 337) says, "que j'appelle système des Monts Notre-Dame" extends from the tip of Gaspé to Lake Champlain and includes shales of the St. Lawrence lowlands as well as the rocks beneath Quebec City. He mentions the difficulty of separating the range from the Green Mountains, which according to him extend northward to the Chaudière River. Logan (1863, pp. 1–3) in his "Geology of Canada" apparently considered that what he termed the Notre-Dame range extends from Vermont through to the eastern tip of Gaspé with the Shickshocks an elevated part of it. Hunt (1878, p. 83E) gives a clear statement to the effect that the Notre-Dame range extends 150 miles from Vermont and is the extension of the Green Mountains. It terminates close to the Etchemin River and south of the Ile d'Orléans, but the Shickshock range is a prolongation of it after an interruption of 250 miles.

Some atlases used in Canadian schools show the Notre-Dame Range extending from Vermont to Gaspé but southeast of the line referred to as the Sutton axis. However, the name Notre-Dame seems to have but little currency. As a schoolboy in British Columbia I learned of the Shickshock and Notre-Dame ranges; since residing in Quebec I cannot recall hearing any geologist mention the Notre-Dame range, although Shickshock is commonly mentioned. My colleagues at Laval have the same experience. In other words the name "Notre-Dame" is not used. Under these circumstances a revised nomenclature for the physiographic feature of the Appalachian region is in order and it is hoped that this symposium can provide some stimulus toward its achievement.

The Use of the Name Appalachian

It is appropriate to examine the first geological map of North America to determine how the Appalachian region as a whole was considered. This map (Fig. 1) is the work of Guettard (Guettard 1756) who, without visiting America, compiled it in 1752 according to his system. In it "Les Apalaches Monts" appear at a locality southeast of a branch of what is now called the Tennessee River in North Carolina and Georgia. The extension of the range is interrupted near the site of Pittsburg by the unit referred to by Guettard as the "Bande marneuse." This band includes limestones and other sedimentary rocks, but Guettard shows it as covering the Adirondacks by extending northward to the east end of Lake Ontario and from thence along the south shore of the St. Lawrence River to the east end of Montreal Island where it crosses to the north side to extend to and along Maskinongé River, where a tongue bounded on the east by St. Maurice River extends northwestward to the height of land which is designated "Montagnes du Nord" in the "Pays des Esquimaux." The band trends east to cross the St. Lawrence River at the upstream end of the Ile d'Orléans and from there runs southeast to the Atlantic ocean. The "band" thus crosses the Appalachian region as now recognized, and there is no evidence, other than appearance of the name "Mt. Louis," that Guettard recognized that "Gaspie" is mountainous.

Philippe Buache, who was later to achieve notoriety through his support, in 1790, of the apocryphal tale of Lorenzo Ferrer de Maldonado's voyage in 1588 through a Northwest passage, was apparently responsible for the cartography on Guettard's map, but on a map published in the same year as Guettard's (Buache 1756) he shows a mountain range not only extending

FIGURE 1. Guettard's map of 1752.

up the east side of the continent but also crossing through St. Georges Bank to Europe. Guettard acknowledges receiving information from La Galissonière, who as governor of New France gave abundant evidence of his awareness of the importance of the continuous Appalachian barrier in containing the westward advance of the New England colonists, but his map does not show a continuous range.

During the next century and a half, there was a tendency to assimilate the Alleghany Mountains to the Appalachians, and, according to different authors, the one can be considered part of the other. Although Bouchette (1815, p. 25), for example, stated that mountain ranges of the Eastern Townships are continuations of those of New England, he refrained from using any names for the ranges. However, by 1832, as stated by Abbé Holmes in "Géographie," "Apalaches (Alleghanys)" mountains extend from Florida to the Gulf of St. Lawrence. In the second and third editions of Dana's textbook, the Appalachian Mountains are said to extend through the Green Mountains to the vicinity of Quebec.

Usage would seem to allow that they extend to the tip of Gaspé, however some dissent from this is apparent. Woodward (1957) in a paper on the direction of the late folding would limit the Appalachians to a region south of the Adirondacks. Again he (Woodward 1957a) states, in speaking of late Palaeozoic orogeny in the New England-Acadian country, "which should not carry the name 'Appalachian' anyway."

Probably only a very few United States geologists would restrict the name Appalachian to the eastern part of the United States of America, for most workers in New England recognize the extension of the tectonic unit into Quebec. Many of the ages assigned to formations there were based on fossils from Quebec. The metamorphic overprint is less severe in the Eastern Townships than it is in Vermont, and, although the area is not rich in fossils, those found in Quebec have been of substantial use in dating.

GEOLOGICAL INTERPRETATION

The present understanding of the geology of the Quebec Appalachian region is essentially similar to that of Logan. In 1843, before he had undertaken his principal task as Director of the Geological Survey, with keen insight and with equal daring he inferred that scattered outcrops of Siluro-Devonian rocks in the Memphremagog district and at St-Georges-de-Beauce might be joined to some in Gaspé as part of an axial belt extending from Vermont to the tip of Gaspé. The axial belt is flanked by older rocks. This interpretation was later presented on a coloured map issued in 1855 to accompany the *Esquisse géologique du Canada* by Logan and Hunt. The large hand-coloured map draughted to accompany the 1863 volume of the geology of Canada shows the Logan interpretation and has a very modern look.

Although Logan's work had been subject to attack by his contemporaries, notably by the strong advocate of the Taconic system, J. Marcou, the most virulent attack was launched by A. R. C. Selwyn (1879, 1833) who became second director of the Geological Survey in 1869. Selwyn considered the axial part of the Eastern Townships to be Precambrian with younger rocks flanking it. This is set forth in the map of 1882 (Geol. Surv. Can., Map 411). On this map the large area of Siluro-Devonian rocks was still shown in the Eastern Townships, but on the Eastern Township map sheets prepared by Ells the Silurian outcrop is reduced, for reasons that are not obvious, to insignificant patches. This was the interpretation accepted on Geological Survey maps from 1913 until 1955 (Maps 914A, 1045A).

By about 1950 mapping had advanced and with more rigorous stratigraphic and structural analyses, a return to the Logan interpretation became mandatory. The structural unit of Siluro-Devonian rocks extending through Gaspé and Vermont was, after consultation with New England geologists, referred to as the Gaspé-Connecticut River synclinorium. The rocks themselves can be referred to as the Gaspé Group, a name used by Logan. The Gaspé Group is composed of formations of diverse lithology. On the Gaspé end of the synclinorium, carbonates and siltstones and shales occur in considerable force, whereas in the Eastern Townships coarser clastic rocks with minor carbonates are the rule; furthermore, the oldest members in Gaspé are Early Silurian whereas the oldest recognized to the southwest are Lower Ludlow.

The re-assignment of some formations to the Gaspé Group tended to reduce somewhat the apparent diversities of the formations considered older than the Gaspé Group. Some fossil localities have been discovered, some stratigraphic units have been traced, and some units have been divided, but there still remains a substantial number of formations or even stratigraphic units of higher order whose relationships to other units or whose correct age is unknown. The difficulty of correlation exists not only along and between segments but also across the segments. Such difficulties arise because of drastic changes in lithofacies, and the resolution of these difficulties will no doubt contribute to the understanding of the geology of the region.

REFERENCES

BOUCHETTE, JOSEPH (1815). Topographical description of the province of Lower Canada with remarks upon Upper Canada, and the relative connexion of both provinces with the United States of America: 1–640, W. Foden, London.

BUACHE, PHILIPPE (1756). Sur les chaînes de montagnes du globe terrestre, *in* Histoire de l'Académie Royale des Sciences, 1752: 117–124.

——— Essai de géographie physique où l'on propose des vues générales sur l'espèce de charpente du globe, composée des chaînes de montagnes qui traversent les mers comme les terres, avec quelques considérations particulières sur les différents bassins de la mer et sur sa configuration intérieure: 399–416.

CLARK, T. H. (1951). New light on Logan's Line. Trans. Roy. Soc. Can. Ser. III, vol. XLV, Sec. IV: 11–22.

GUETTARD, J. E. (1756). Sur la comparison du Canada avec la Suisse, par rapport à ses minéraux, *In* Histoire de l'Académie Royale des Sciences, 1752: 12–16, 323–360, 524–538.

HUNT, T. S. (1878). Special report on the trap dykes and Azoic rocks of southeastern Pennsylvania; Second Geol. Surv. Pennsylvania, 1875: *1*, 1–253E, Harrisburg, Pa.

LOGAN, W. E. and HUNT, T. S. (1855). Esquisse géologique du Canada pour servir à l'intelligence de la carte géologique et la collection des minéraux économiques envoyées à l'Exposition Universelle de Paris 1855.

MARCOU, JULES (1955). Esquisse d'une classification des chaînes de montagnes d'une partie de l'Amérique du Nord. Annales des Mines: *VII*, 329–350, Paris.

TRUDEL, M. (1961). Atlas historique du Canada français. Québec: les presses universitaires Laval.

WOODWARD, H. P. (1957). Structural elements of northeastern Appalachians. Am. Assoc. Pet. Geol. Bull. 47: 1428–1440.

——— (1957a) Chronology of Appalachian folding. Am. Assoc. Pet. Geol. Bull. 47: 2312–2327.

THE TACONIC UNCONFORMITY IN THE GASPÉ PENINSULA AND NEIGHBOURING REGIONS*

W. B. Skidmore

ABSTRACT

The Gaspé-Connecticut Valley synclinorium forms, at its northeast end, the central Silurian-Devonian belt of the Gaspé Peninsula, bordered on both sides by Cambro-Ordovician rocks. Although only about 40 miles apart, the bordering regions have had very different tectonic and sedimentary histories. A Taconic unconformity is clearly present along the north side, but has never been certainly established on the south side. Whereas on the north there is a gap in the sedimentary record embracing at least the whole of the Upper Ordovician, on the south accumulating palaeontological evidence suggests generally continuous sedimentation from Middle Ordovician to Early Devonian, the gap being filled by limestones of Late Ordovician to Silurian age, which can be traced along an anticlinorial belt southwestward into northeastern Maine. Southeast of the anticlinorium, near Chaleurs Bay, an Upper Ordovician unconformity is again evident.

THIS PAPER is not concerned with tectonics as revealed through the study of geological structures, but rather with an aspect of tectonic history suggested by regional stratigraphy, and in particular by the gaps in the stratigraphic record. It is necessarily based on the findings of many workers, and presentation of detailed evidence for the conclusions drawn would be impractical. Data are derived from two main sources: published reports on the region and the, so far, unpublished results of recent workers. The sources of many specific items of information are mentioned in the appropriate places, but a general acknowledgement is here made to all those whose work is briefly summarized.

THE CENTRAL SYNCLINORIUM

It is now an accepted fact that a continuous belt of Devonian and Silurian rocks, flanked on either side by older rocks, extends through the Northern Appalachians from Connecticut to the tip of the Gaspé Peninsula. In the Gaspé Peninsula this feature is generally known as the central Silurian-Devonian belt.

The limit of the synclinorium on its northwest side is easily defined, owing

*Published by permission of the Deputy Minister, Quebec Department of Natural Resources.

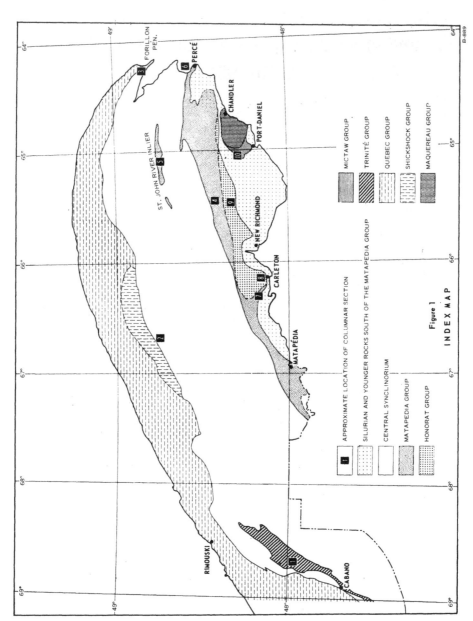

Figure 1

INDEX MAP

FIGURE 1

to the coincidence of lithologic, faunal, and structural discontinuities along the south edge of the Quebec and Shickshock groups. On the southeast side there is less certainty, for reasons to be mentioned later, but here the limit can most conveniently be placed along the north edge of the Matapédia Group.* Thus defined, the synclinorium is about 20 miles wide at the east end of the Peninsula, and widens westward to about 55 miles in the Rimouski region. There are two inliers, that of the Trinité Group in the Rimouski region, and one exposing part of the Matapédia Group along the Saint John River anticline. The synclinorium contains strata ranging in age from Earlier Llandoverian through the Early Devonian, possibly into the Middle Devonian. The lithology of these strata varies greatly, both up-section and laterally, but nevertheless they are as a whole easily distinguished from the rocks on either flank and in the inliers. There are no well-established unconformities in the sequence. The intensity of deformation is variable, increasing in general to the south, but all folding is attributed to the Acadian revolution of approximately Late Devonian time.

Although the oldest strata in the synclinorium are of Early Llandoverian age, it is by no means true that the basal strata are everywhere so old. Nor is it true, according to recent evidence, that the underlying, extrasynclinorial rocks are everywhere older. The contact relations can most conveniently be described under two separate headings, one covering the north and northwest flank of the synclinorium, together with the Trinité inlier, the other covering the south and southeast flank and the Saint John River inlier.

(a) The North Flank

Southwest of the Gaspé Peninsula proper, in the Témiscouata region, the basal synclinorial unit is the Lower Llandoverian Cabano Formation (Lespérance and Greiner, in press). On the Forillon Peninsula, at the east end of Gaspé, according to the latest information (Boucot 1965, p. 2298), the Silurian is absent, and the lowest Devonian beds lie directly on the Québec Group. The basal strata have not been continuously traced between these two extremities, but where they are known they are generally dated as Upper Llandoverian to Wenlockian (e.g. Béland 1960, p. 3; Ollerenshaw 1963, p. 207). It is clear then that the basal beds are, generally, younger to the northeast than to the southwest, and that their age range covers the whole Silurian Period. The contact itself, although faulted along parts of its length and exposed in only a few places, is well established as an angular unconformity (Cumming 1959, p. 15; Lajoie 1962, p. 7).

The underlying rocks of the Quebec and Shickshock groups are structurally complex, and their stratigraphy and age relations are not known in any detail. However, it appears that the rocks close to the contact are mainly Cambrian to Lower Ordovician (Ollerenshaw 1963, pp. 24–34). In the

---

*The term "Matapédia Group" as used in this paper includes rocks mapped as the White Head and Pabos formations, but excludes rocks assigned to the Honorat Group.

eastern part of the region they are flanked on the north by Middle Ordovician rocks (McGerrigle 1953). A small fault slice of Middle Ordovician rocks is found along the south edge of the Shickshock Group (Mattinson 1964, p. 69), and the Trinité Group is dated as Middle Ordovician (Lajoie 1962, p. 4). The Upper Ordovician seems to be entirely absent from this northern flank region. It is not yet known whether there are any unconformities within the Cambro-Ordovician sequence.

Thus the period of zero net deposition (and presumably of zero net subsidence) along the north side of the synclinorium varies from a maximum including the whole Silurian and most or all of the Ordovician at Forillon Peninsula, to a minimum of the Late Ordovician in the Témiscouata region.

(b) The South Flank

Along the south edge of the synclinorium the situation is less clear. At the eastern end of the belt, near Percé, the oldest rocks above the White Head Formation are again apparently of earliest Devonian age (Boucot, personal communication; Copeland 1963), but the contact is faulted, and intermediate Upper Silurian beds might have been cut out. To the west the basal synclinorial rocks are not well dated. The few accurately dated localities are of Ludlovian age (Boucot, personal communication, 1964), but older rocks may be present. On the Saint John River inlier, about ten miles north of the south boundary, White Head limestones are overlain by rocks as old as Upper Llandoverian (Cumming 1959, pp. 25–27).

The nature of the contact itself is also less clear than on the north side. Without entering into a detailed argument on this subject it can be said that there is no unequivocal evidence for an angular unconformity, although there is some local evidence for erosional disconformity at about this level. Workers who in the past have postulated a major unconformity have done so in the belief that the Taconic unconformity recognized to the north should be present along this contact also. As will be seen in a moment, this expectation no longer seems to be justified.

The underlying Matapédia Group, together with the Honorat Group, occupies an anticlinorial belt up to 15 miles wide that runs southwestward from Percé through the length of the peninsula, and continues across northern New Brunswick into northeastern Maine (Pavlides and Berry 1966, p. B53). Its internal relations are not well understood. The Matapédia Group has for many years been dated as Late Ordovician on the basis of fossils collected from a few localities in the Metapédia valley and at Percé (Alcock 1935, pp. 18–26). More recent work by Lespérance (1963–66), Copeland (1964, 1966), Cumming (1963, 1965), and Wright (Boucot, personal communication, 1965) on fossil collections from the eastern part of the Peninsula has resulted in assigned ages for individual collections ranging from Clintonian, through Late and Middle Ashgillian to Caradocian. There is no direct evidence for any break in the sequence. Its base is exposed only near Percé, where it overlies, with apparent angular unconformity, the Upper and Middle Cambrian Murphy Creek and Corner of the Beach formations.

The Silurian collections all come from the White Head Formation within about a 40-mile radius from Percé. One locality is on the Saint John River inlier, the others are all lying close to the south edge of the synclinorium. They indicate that the Upper White Head is younger than the Cabano

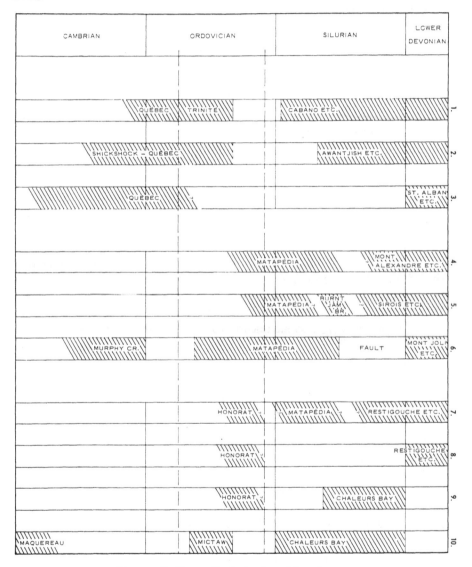

FIGURE 2. Generalized stratigraphic columns.
North flank of synclinorium: 1. Témiscouata area, 2. North-central Gaspé, 3. Forillon Peninsula; *South flank of synclinorium*: 4. Mount Alexandre area, 5. Saint John River, 6. Percé; *Chaleurs Bay Region*: 7. West of Carleton, 8. Carleton, 9. Honorat area, 10. Weir area.

Formation, and that therefore there is an age overlap between the synclinorial and extrasynclinorial rocks. Thus, if there is an unconformity above the White Head, it cannot be correlated with that which passes beneath the Cabano. Moreover, there seems no reason to believe that there is any significant time gap in the stratigraphic sequence at this level, although, except on the Saint John River inlier, the presence of the Wenlock remains to be definitely established.

It is interesting to note that Pavlides (Pavlides and Berry 1966) has discovered evidence for a continuous sequence from Middle Ordovician to Upper Silurian at the other end of the anticlinorium in northeastern Maine.

CHALEURS BAY REGION

The Matapédia and Honorat groups are overlain again on the south by Silurian and Devonian rocks generally similar to those within the synclinorium (Chaleurs Bay and Restigouche groups), but the contact relations here are different from those along the north side of the anticlinorium. From Chandler westward to Carleton the southern part of the anticlinorium is occupied by the Honorat Group (Ayrton 1964; Skidmore 1965). The rocks of this group have previously been considered a part of the Matapédia Group, but they are lithologically quite distinct. Their relation to the Matapédia Group is still somewhat obscure, mainly because of extensive faulting along the contact. They may be a facies equivalent of the older part of the Matapédia. Palaeontologic evidence is so far too scanty to settle this question.

To the east of New Richmond the Honorat is probably, though not certainly, overlain with angular unconformity by Upper Llandoverian rocks at the base of the Chaleurs Bay Group. Westward from New Richmond, in the neighbourhood of Carleton, the Silurian formations of the Chaleurs Bay Group are missing, and the Honorat is overlain, here with abrupt angular unconformity, by Lower Devonian strata. Westward again from Carleton, three changes take place within a short distance: (1) Upper (and then Middle) Silurian rocks of the Chaleurs Bay type reappear below the Lower Devonian; (2) rocks similar to the Upper White Head (though so far undated) appear between these and the Honorat Group; (3) the angular unconformity is no longer evident.

It appears then that this southern edge of the Matapédia anticlinorium is an unconformity, although structural discordance and the stratigraphic time gap vary along its length. The age of the unconformity must be pre-Late Llandoverian, and is very probably post-Middle Ordovician.

A few miles again to the southeast from this contact, in the Port-Daniel area, the Chaleurs Bay Group overlies the Mictaw and Maquereau groups. The Mictaw Group is Middle Ordovician, and the Maquereau is still older, probably Cambrian or Precambrian. Although the Chaleurs Bay is not clearly seen in outcrop to rest with angular unconformity on the Mictaw, the fact that it cuts across the Mictaw to rest on the Maquereau is strongly suggestive of structural discordance. The basal strata of the Chaleurs Bay are dated as Upper Llandoverian near the coast and as Lower Llandoverian inland, in southeastern Weir Township (Ayrton 1964). In this district we have therefore, as in the Témiscouata region, a minimum period of zero net deposition encompassing the Late Ordovician.

Summary and Outstanding Problems

Along the north edge of the central synclinorium, and again, some 50 miles to the southeast, on Chaleurs Bay, there is a clear unconformity between Cambro-Ordovician and Siluro-Devonian rocks. In both regions the stratigraphic time gap, in so far as it has been dated, covers at least the whole of the Upper Ordovician. While it cannot be certain that a single Late Ordovician phase of uplift was responsible for the whole stratigraphic gap everywhere in these regions, it is at least unnecessary to postulate any other event. It has thus been natural to attempt to trace the same unconformity across the intervening space, and equally natural to expect that it should be found between the Ordovician Matapédia Group and the overlying Silurian rocks of the synclinorium. Recent evidence for the Silurian age of the uppermost part of the Matapédia Group suggests strongly that any unconformity at this level must be relatively minor, and in any case cannot be correlated with the Upper Ordovician unconformity. This evidence, together with the presence of a thick section of Upper Ordovician rocks, and some Middle Ordovician rocks, in the Matapédia Group, leads one to suppose that the Taconic unconformity is simply not present in this intervening region, and that the Matapédia Group represents a trough where sedimentation continued while uplift took place on both flanks.

However, it must be admitted that the unconformity, albeit represented by a relatively short time-stratigraphic gap, might be present at a lower level in the section, for instance between the Upper and Middle Ordovician parts of the Matapédia, or between the Honorat and Matapédia groups. The question will not be settled without a better understanding of the internal stratigraphy and structure of the older part of the Matapédia Group, and of the age and relations of the Honorat. At present there is conflicting evidence on the possibility of structural discordance between the two groups (Ayrton 1964, pp. 84–86).

In any case, it already seems clear that the effect of the Taconic orogeny was very varied even within this comparatively small region. A more detailed knowledge of the variation in the time-stratigraphic gap, and in the amount of structural discordance, may eventually lead to a fairly complete picture of the tectonic pattern of that time.

The apparent absence of Lower Ordovician rocks in southern Gaspé and the unconformities between the Murphy Creek and White Head formations, and between the Maquereau and Mictaw groups, suggest an earlier and quite separate episode of uplift in the Early Ordovician, the extent of which has not been established. A less apparent Early Ordovician unconformity could well be present to the north of the synclinorium. Its discovery would depend upon further study of the stratigraphy of the Quebec Group.

REFERENCES

ALCOCK, F. J. (1935). Geology of Chaleur Bay region. Geol. Surv. Can., Mem. 183.

AYRTON, W. G. (1964). A structural study of the Chandler-Port-Daniel area, Gaspé Peninsula, Quebec. Ph.D. dissertation, Northwestern University.

BÉLAND, J. (1960). Preliminary report on Rimouski-Matapédia area, electoral districts of Rimouski, Matapédia, Bonaventure and Matane. Que. Dept. Mines, P. R. 430.

BOUCOT, A. J. (1965). Silurian stratigraphy of Gaspé Peninsula, Quebec. Bull. Am. Assoc. Pet. Geol., *49* (12): 2295–2303.

COPELAND, M. J. (1963, 1964, 1966). Geol. Surv. Can., manuscript paleontological reports.

CUMMING, L. M. (1959). Silurian and Lower Devonian formations in the eastern part of Gaspé Peninsula, Quebec. Geol. Surv. Can., Mem. 304.

——— (1963, 1965). Geol. Surv. Can., manuscript paleontological reports.

LAJOIE, J. (1962). Preliminary report on Chénier-Bédard area, Rimouski county. Que. Dept. Nat. Res., P.R. 493.

LESPÉRANCE, P. J. (1963–66). Que. Dept. Nat. Res., manuscript paleontological reports.

LESPÉRANCE, P. J. and GREINER, H. R. (in press). Squateck-Cabano area, Rimouski, Rivière-du-Loup and Témiscouata counties. Que. Dept. Nat. Res., G. R. 128.

MATTINSON, G. R. (1964). Mount Logan area, Matane and Gaspé-North counties. Que. Dept. Nat. Res., G.R. 118.

McGERRIGLE, H. W. (1953). Geological map of Gaspé Peninsula. Que. Dept. Mines, Map 1000.

OLLERENSHAW, N. C. (1963). Stratigraphic problems of the western Shickshock Mountains (Cuoq-Langis area). Ph.D. thesis, University of Toronto.

PAVLIDES, L. and BERRY, W. B. N. (1966). Graptolite-bearing Silurian rocks of the Houlton-Smyrna Mills area, Aroostook county, Maine. U.S. Geol. Surv., Prof. paper 550-B: B51–B61.

SKIDMORE, W. B. (1965). Honorat-Reboul area, Bonaventure county. Que. Dept. Nat. Res., G.R. 107.

TECTONICS OF PART OF THE SILLERY FORMATION IN THE CHAUDIÈRE-MATAPÉDIA SEGMENT OF THE QUÉBEC APPALACHIANS*

Claude Hubert

ABSTRACT

On the south shore of the St. Lawrence, between Montmagny and Rivière-du-Loup, the Cambro-Ordovician Sillery Formation constitutes a complex synclinorium which extends to and flanks the northwestern side of the Sutton anticlinorium. The synclinorium is broken by several *en echelon* reverse faults and is composed of imbricated wedges and blocks, brought into contact from different sites of origin across the basin of deposition.

THE CHAUDIÈRE-MATAPÉDIA segment of the Québec Appalachians is divided into two broad zones: the Taconic folded belt on the northwest and the Acadian on the southeast (Fig. 1). The Taconic folded belt can be subdivided further into three subzones on a stratigraphic-structural basis. These three subzones are: (1) the Sillery belt along the St. Lawrence River, (2) the Rosaire-Caldwell belt characterized by refolded structures, and (3) the Beauceville belt to the south.

On the south shore of the St. Lawrence, between Montmagny and Kamouraska, the Cambro-Ordovician Sillery belt is characterized by a series of imbricated litho-structural slices or blocks which are, from northwest to southeast, the St-Roch, La Pocatière I, La Pocatière II, and Armagh (Fig. 2). Massive and thick-bedded feldspathic sandstones constitute the Armagh Group. Rocks of the St-Roch Formation are chiefly mudstones with intercalated bands of conglomerates, feldspathic sandstones, siltstones, shales, and limestones. The St-Damase Formation is composed of feldspathic sandstones in thick, graded beds with lenticular bands of orthoquartzite and polymictic limestone conglomerate. The Kamouraska Formation is made up of orthoquartzite and polymictic limestone conglomerate. Shales and siltstones constitute the bulk of the Rivière Ouelle Formation. Table I shows the succession of the formations in each slice and their correlations. In this area, the Cambro-Ordovician Sillery shows a marked lithological change from southeast to northwest and, correspondingly, from sandstones to shales and mudstones. Several thin bands of limestone are also present to the northwest and these are absent in the Armagh.

*Published with the permission of the Deputy Minister, Department of Natural Resources, Québec.

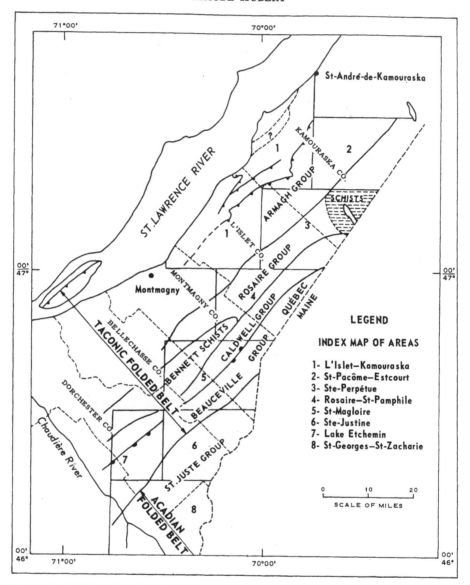

FIGURE 1. Outline of regional geology between the Chaudière River and Kamouraska County (modified from Béland 1962).

In each of these slices, the rocks are folded into a series of *en echelon*, doubly-plunging, narrow, steep anticlines and broad, shallow synclines which trend northeast-southwest. The folds are overturned to the northwest and generally have steeper plunges towards the northeast (from 5 to 25 degrees) than the southwest (from 2 to 10 degrees). Second-order parasitic folds are superimposed on their limbs. They are also overturned to the northwest but are conspicuously inequant with one limb much longer than the

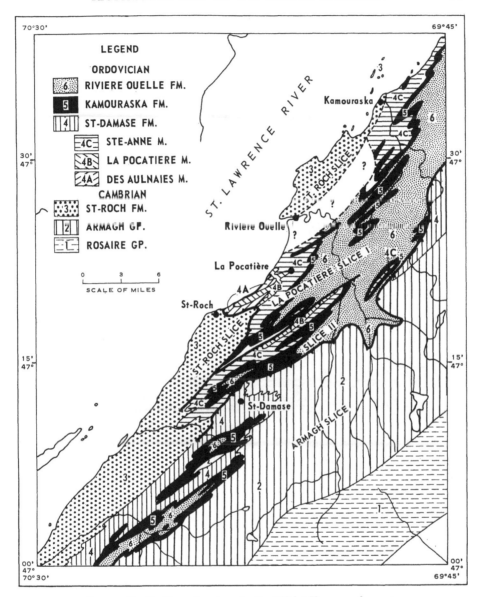

FIGURE 2.   Outline of geology in the L'Islet-Kamouraska area.

other. Their fold axes are either parallel to the general trend of the major fold or cut them at an angle.

In the Armagh slice, the internal structure is not well established owing to the lack of marker beds. Jacques Béland and the writer think that the Armagh is folded into a broad anticline with minor parasitic folds on the limbs.

Three reverse faults are necessary to explain the imbrication and repetition

TABLE I

Homotaxial Correlation of Formations in the L'Islet-Kamouraska Area. (Rosaire Group not included)

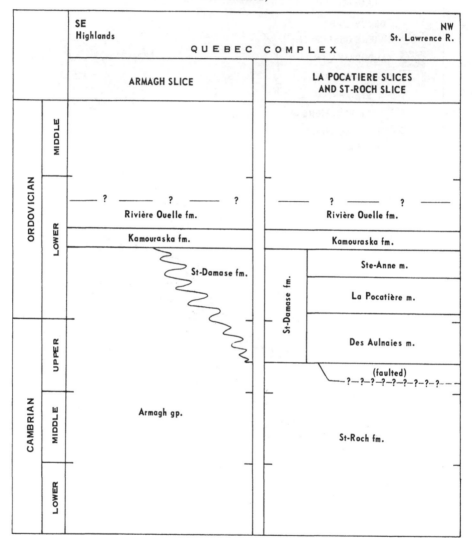

of litho-structural blocks in the Montmagny-Kamouraska area. The three faults separate, respectively, the St-Roch and La Pocatière I slices, the La Pocatière I and La Pocatière II slices, the La Pocatière II and Armagh slices. The St-Roch is probably also bounded on the northwest by a fault. This postulated fault separates the Middle Ordovician rocks on Goose Island on the St. Lawrence River from the Lower Cambrian rocks of the St-Roch Formation on the south shore of the river.

Two aspects of the tectonics of the Sillery remain to be discussed. These are its tectonic framework and its tectonic evolution through the Cambrian and the Ordovician periods. Stratigraphic and lithological observations as well as paleocurrent data suggest that the various units of the Sillery probably formed a group of facies in the original basin of deposition. From northwest to southeast this series consisted of a Cambrian sequence of quartz sandstone and limestone formation, which passed in a southeasterly direction into the mudstones of the St-Roch Formation, and the graded-bedded arkoses of the St-Damase Formation (overlying the St-Roch) which, in turn, passed in a southeasterly direction into the massive and thick-bedded arkoses of the Armagh Group. The width of each of the lithosomes is unknown, thus the four palinspastic block diagrams (see Fig. 3 A–D) are diagrammatic only.

The arkosic material which constitutes the Armagh, St-Roch, and St-Damase units was derived from islands and/or a landmass located to the southeast. This source area was a landmass of high relief and similar in texture and composition to rocks of the Precambrian Shield. To the northwest, the textural and mineralogical maturity of the quartz which now constitutes the Kamouraska quartzites would indicate that the Canadian Shield was a landmass of very low relief.

At the beginning, the basin consisted possibly of a stable shelf to the northwest and an unstable aggradation shelf to the southeast. In early Upper Cambrian time, a sufficient difference in elevation must have existed between the unstable shelf to the southeast and the shale environment immediately to the northwest because the rocks overlying the St-Roch mudstones are the product of cannibalistic sedimentation. Almost all of the beds of the St-Damase Formation show excellent and repetitive graded bedding.

Crustal instability is also indicated on the northwestern side of the basin in middle Upper Cambrian time. Parts or all of the stable shelf with the pure quartz sandstone and limestone were uplifted above the sea. Polymictic limestone conglomerates and orthoquartzites were derived from this new uplift and were deposited in the feldspathic sandstone environment. These beds constitute the La Pocatière Member of the St-Damase and are lenticular. This uplift, however, must have been short-lived because the rocks overlying the La Pocatière Member indicate an abrupt return to the conditions of sedimentation which existed prior to the uplift.

In latest Upper Cambrian and/or Early Ordovician time, a second major tectonic pulse uplifted the quartz sandstone and limestone shelf on the northwest. The blanket deposit produced is the Kamouraska Formation. The development of this formation marked a major change in the regime of sedimentation of the Sillery in this area. The thin-bedded shales and siltstones of the Rivière Ouelle Formation show very marked structural and textural changes from all of the coarse-grained and locally derived formations of Cambrian age which underlie the Kamouraska Formation.

In detail, the structure and stratigraphy of the Sillery is complex and

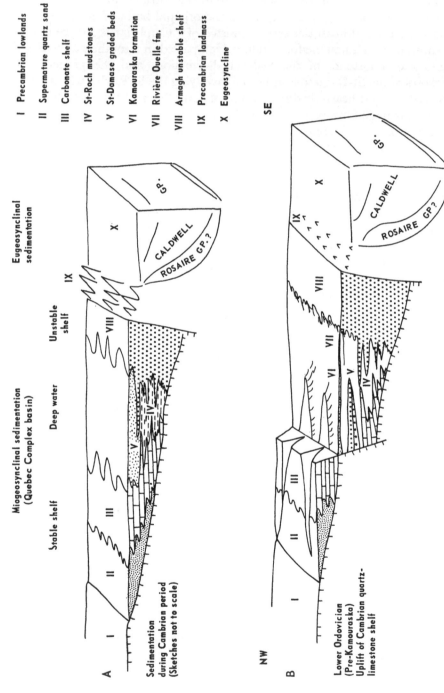

APPALACHIAN TROUGH

Miogeosynclinal sedimentation
(Quebec Complex basin)

Eugeosynclinal
sedimentation

Stable shelf   Deep water   Unstable shelf

I   Precambrian lowlands
II   Supermature quartz sand
III Carbonate shelf
IV   St-Roch mudstones
V   St-Damase graded beds
VI   Kamouraska formation
VII Rivière Ouelle fm.
VIII Armagh unstable shelf
IX   Precambrian landmass
X   Eugeosyncline

Sedimentation
during Cambrian period
(Sketches not to scale)

Lower Ordovician
(Pre-Kamouraska)
Uplift of Cambrian quartz-
limestone shelf

FIGURE 3. Diagrammatic cross-sections of the Quebec Complex during sedimentation and after folding.

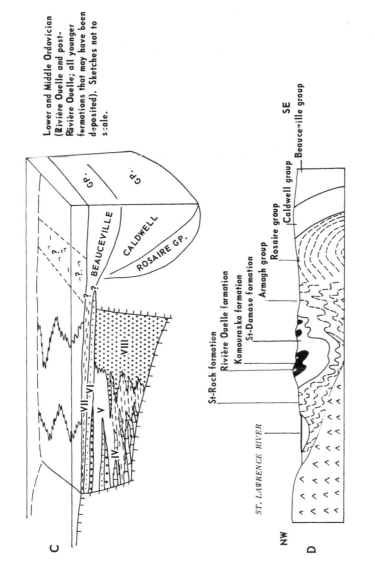

Lower and Middle Ordovician (Rivière Ouelle and post-Rivière Ouelle; all younger formations that may have been deposited). Sketches not to scale.

FIGURE 3 (*continued*)

difficult to map, but on a regional scale the Sillery Formation shows consistent stratigraphic relationships between its units and its structure appears to be a complexly folded synclinorium which is broken up by several local internal faults.

REFERENCES

BÉLAND, J. (1957). St-Magloire and Rosaire-St-Pamphile Areas. Que. Dept. Mines G.R. 76; 49 pp.

———— (1962). Ste-Perpétue Area. Que. Dept. Nat. Res., G.R. 98: 22 pp.

GORMAN, W. A. (1954). The Ste-Justine Map-Area. Que. Dept. Mines, P.R. 297.

———— (1955).The St-Georges-St-Zacharie Map-area. Que. Dept. Mines, P.R. 314.

———— (1956). The St-Pacôme-Estcourt Area. Que. Dept. Mines, unpub. ms. 29 pp.

HUBERT, C. (1965). Stratigraphy of the Quebec Complex in the L'Islet-Kamouraska Area, Quebec. Unpublished Ph.D. Thesis, McGill University.

TOLMAN, C. (1936). Lake Etchemin Map-Area, Quebec. Geol. Surv. Can., Mem. 199.

TECTONICS OF PART OF THE APPALACHIAN REGION OF SOUTHEASTERN QUÉBEC (SOUTHWEST OF THE CHAUDIÈRE RIVER)*

Pierre St-Julien

ABSTRACT

The southeastern part of the Québec Appalachians is underlain by Cambro-Devonian rocks arranged in five northeasterly trending belts. These are, from northwest to southeast, the Sutton-Bennett belt, the Serpentine belt, the St-Victor synclinorium, the Stoke Mountain belt, and the Gaspé-Connecticut Valley synclinorium.

The Sutton-Bennett belt is characterized by refolded recumbent structures developed in schists presumed to be Cambrian.

The folds of the Serpentine belt have steeply dipping limbs and steeply plunging axes. In ascending order, the rock units involved are part of the Rosaire Group, the Caldwell-Mansonville formations, and the St-Daniel and Brompton formations. All three units are presumed to be Cambrian to Lower Ordovician.

The St-Victor synclinorium is characterized by gently plunging asymmetric folds and includes the Beauceville, St-Victor, and Sherbrooke formations of Middle to Upper Ordovician ages. It also includes some Upper Silurian lying unconformably on the Ordovician.

The Stoke Mountain belt includes the Ascot and the Weedon Schist formations of Cambrian to Lower Ordovician age. The major folds of the Stoke Mountain belt are isoclinal and strongly overturned to the northwest. Later movement produced open folds trending northeast-southwest.

The Gaspé-Connecticut Valley synclinorium exhibits tight, gently plunging folds generally overturned to the southeast. It consists of the Siluro-Devonian St-Francis Group, which overlies unconformably the Cambro-Lower Ordovician schists of the Stoke Mountain belt. This synclinorium and the Stoke Mountain belt have been thrust northwestward over the St-Victor synclinorium.

THE SOUTHEASTERN PART of the Appalachian region of Québec southwest of the Chaudière river consists of five northeasterly trending belts. These divisions rest largely on field studies made during the last twenty years and are based mainly on tectonics but also on rock types and ages. The rocks concerned range from Cambrian to Middle Devonian in age and are mainly metasedimentary; metavolcanic and intrusive types are common but less widespread. The five belts are, from northwest to southeast (Fig. 1): (1) The Sutton-Bennett belt, (2) The Serpentine belt, (3) The St-Victor synclinorium, (4) The Stoke Mountain belt, (5) The Gaspé-Connecticut Valley synclinorium.

The Sutton-Bennett, the Serpentine, and the Stoke Mountain belts include

*Published by permission of the Deputy Minister, Québec Department of Natural Resources.

rocks presumed to be Cambrian to Lower Ordovician in age. Rocks of the St-Victor synclinorium are Middle and Upper Ordovician and Upper Silurian, and those of the Gaspé-Connecticut Valley synclinorium are Silurian and Devonian.

Ultramafics and meta-gabbros are the most common intrusive rocks in the Serpentine and in the Sutton-Bennett belts. Albite granite and albite rhyolite porphyry are common in the Stoke Mountain belt. Dykes of diorite cut rocks of the St-Victor synclinorium. Several massifs of granite are exposed in the Gaspé-Connecticut Valley trough of Siluro-Devonian rocks.

DESCRIPTION OF THE STRUCTURES

(1) *The Sutton-Bennett belt*

This belt is 10 to 12 miles wide and is underlain by schists. It is characterized by two and possibly three different fold patterns or styles: one or two systems of early folds and a superposed system of late folds.

Many geologists working in the area underlain by the Sutton-Green Mountain rocks suggest that at least two distinct phases of deformation were responsible for the essentially east-west or southeast-northwest and north-south or northeast-southwest fold axes (Osberg 1952; Chidester 1953; Albee 1957; Eric and Dennis 1958; De Römer 1960; Eakins 1963; Osberg 1965; Rickard 1965; Roth 1965). Osberg's study (1965) in the Knowlton-Richmond area suggests one and possibly two phases of early deformation: an early northeast fold and an early northwest fold. In the Bennett Schist, near Thetford Mines, we have evidence of a system of recumbent early folds

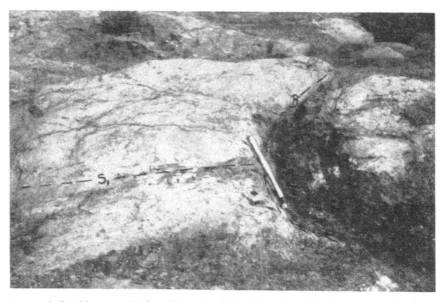

FIGURE 2. Looking east, half a mile north of East Broughton Station: gently plunging, southeast fold (early) in the orthoquartzite of the Bennett Schists. The schistosity (S_1 surface) is nearly horizontal.

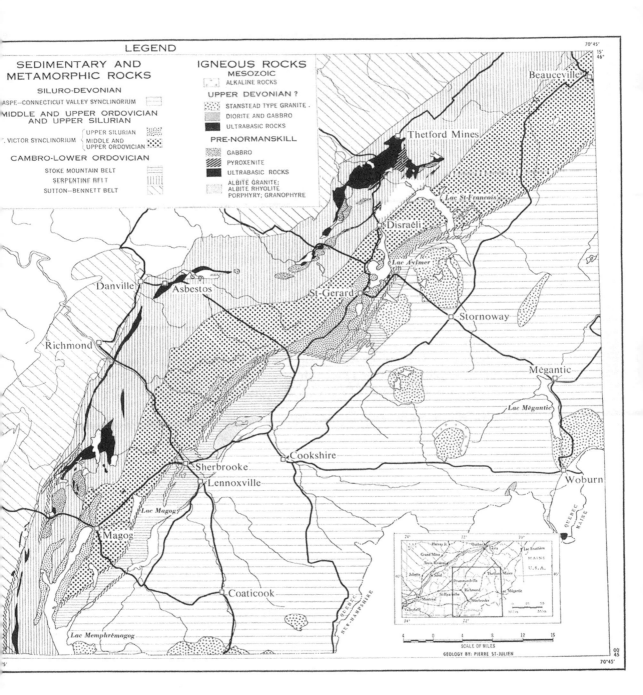

LEGEND

SEDIMENTARY AND
METAMORPHIC ROCKS

SILURO-DEVONIAN

ASPE—CONNECTICUT VALLEY SYNCLINORIUM

MIDDLE AND UPPER ORDOVICIAN
AND UPPER SILURIAN

VICTOR SYNCLINORIUM { UPPER SILURIAN
MIDDLE AND
UPPER ORDOVICIAN

CAMBRO-LOWER ORDOVICIAN

STOKE MOUNTAIN BELT
SERPENTINE BELT
SUTTON—BENNETT BELT

IGNEOUS ROCKS

MESOZOIC
ALKALINE ROCKS

UPPER DEVONIAN ?
STANSTEAD TYPE GRANITE
DIORITE AND GABBRO
ULTRABASIC ROCKS

PRE-NORMANSKILL
GABBRO
PYROXENITE
ULTRABASIC ROCKS
ALBITE GRANITE;
ALBITE RHYOLITE
PORPHYRY; GRANOPHYRE

Beauceville

Thetford Mines

Lac St-Francois

Disraéli

Lac Aylmer

St-Gerard

Stornoway

Danville Asbestos

Mégantic

Richmond

Lac Mégantic

Cookshire

Sherbrooke

Woburn

Lennoxville

Lac Magog

Magog

QUEBEC
MAINE

Coaticook

NEW HAMPSHIRE

Lac Memphrémagog

SCALE OF MILES

GEOLOGY BY: PIERRE ST-JULIEN

FIGURE 1

FIGURE 3. Looking west, 1,500 feet, 320° from Leeds Station: gently plunging southeast fold (early) in the hanging wall of the Pennington Dyke.

(Figs. 2 and 3) with a northwest-southeast axial trace. Superposed on the early folds is the broad arch commonly referred to as the Sutton-Green Mountain anticlinorium. Its axial plane cleavage is nearly vertical. Figure 4 shows in a simplified way the relationships of the early and late folds. Figures 2 and 3 show field examples of some recumbent folds.

2. The Serpentine belt

The Serpentine belt is about eight miles wide and trends northeast-southwest. It is underlain by part of the orthoquartzite-shale assemblage of the Rosaire Group, of the dirty sandstone and volcanic assemblage of the Caldwell-Mansonville formations, and of a slate and pyroclastic assemblage of the St-Daniel-Brompton formations.

In this belt, two systems of early folds are definite. They range in size from minute plications to folds with amplitudes of thousands of feet. The vertical axial plane of the earlier system strikes east-west or northwest-southeast. The axes plunge 45° to 85° east or southeast. The axes of superimposed early folds trend north-south or northeast-southwest and plunge 30° to 90° in these directions. The axial planes of these folds, many of which are isoclinal, are vertical or slightly overturned to the west or northwest.

The early structures of the Serpentine belt appear to be little affected by the late deformation except for the development of a new cleavage and of a few drag folds parallel to the regional trend of the late major arch, the Sutton-Green Mountain anticlinorium.

It is interesting to note that these important masses of ultramafic rocks appear to be located at the intersection of the two early-fold systems and in

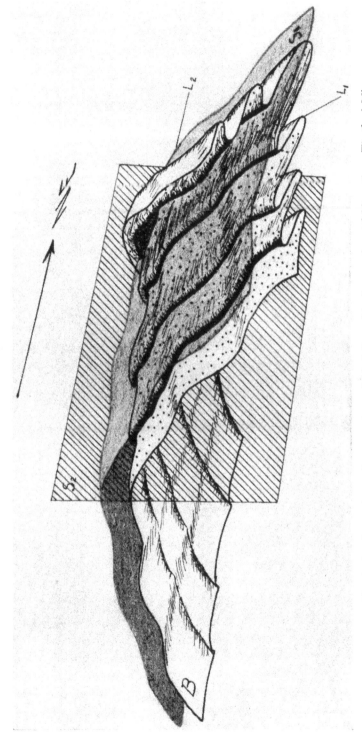

FIGURE 4. Sketch showing early northwest-southeast folds deformed by late northeast fold north or Thetford Mines.

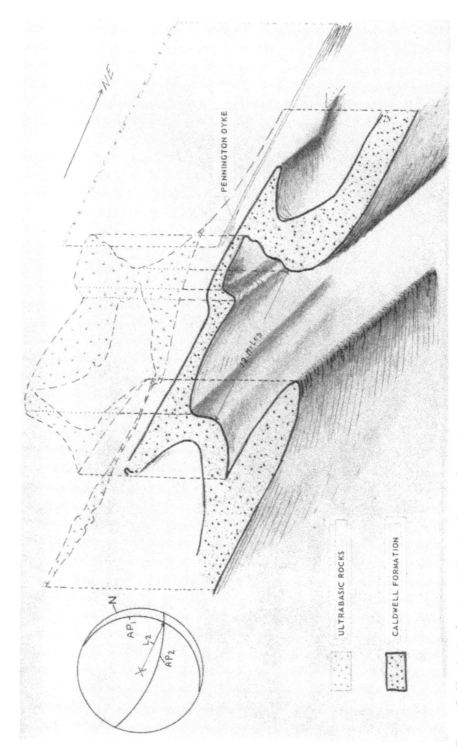

FIGURE 5. Sketch showing the relationships between intersecting early northwest fold and early northeast fold in the Thetford Mines–Black Lake area. The important mass of ultramafic rocks appears to be located at the intersection of these two early fold systems and in a major fault zone. The geometric relationship of these early folds is shown in the cyclogram.

the major fault zones. Figure 5 shows the structural relations between the Caldwell Formation and the ultramafic pipe of Thetford Mines.

3. *The Stoke Mountain belt*

The Stoke Mountain belt includes the Ascot Formation and its correlative, the Weedon Schist Formation, both of which consist of an acidic pyroclastic and shale assemblage.

The same type of deformation is observed in the Stoke Mountain belt as in the Bennett Schist. The early folds are isoclinal and strongly overturned to the northwest; most axes plunge southeast but some plunge northeast or southwest. Both limbs of the early, isoclinal folds are marked by major open sinistral drag folds (Fig. 6). These younger folds have axial surfaces that strike N.53° E. and dip 52° N.W.; their axes trend northeast and have an average plunge of 10 to 40 degrees in this direction.

4. *The St-Victor synclinorium*

The St-Victor synclinorium includes the Beauceville, St-Victor, and Sherbrooke formations of Middle and Upper Ordovician age. It also includes

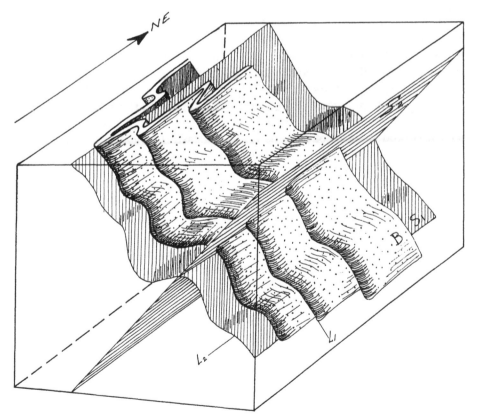

FIGURE 6. Sketch showing the styles of folding observed in the Stoke Mountain belt in the Sherbrooke area.

some Upper Silurian rocks (notably the Glenbrooke, Aylmer Lake, and Cranbourne formations) which lie unconformably on the Ordovician.

The Beauceville lies unconformably on parts of the Cambro-Lower Ordovician rocks of the Serpentine belt and marks the northwestern boundary of the St-Victor synclinorium. The southeastern limit of this belt is a fault.

The Beauceville Formation consists of alternating beds of acidic tuff and graphitic slate. The St-Victor and Sherbrooke formations consist largely of graded beds of graywacke and tuff with interbedded slate. The Upper Silurian formations are dominantly composed of limestone and shale.

Flexural-shear folds are characteristic of the St-Victor synclinorium strata. These form a series of relatively open folds trending northeasterly, with the axial planes vertical or slightly overturned to the northwest. Analysis of field data shows that these folds have a shallow plunge to the southwest near St-Victor-de-Beauce, and to the northeast near Sherbrooke.

5. The Gaspé-Connecticut Valley synclinorium

The Gaspé-Connecticut Valley synclinorium consists of the Siluro-Devonian St. Francis Group, which overlies unconformably the Cambro-Lower Ordovician strata of the Stoke Mountain belt. The rocks of the Gaspé-Connecticut Valley synclinorium and of the underlying Stoke Mountain belt have been thrust northwestward over the St-Victor synclinorium to such degree that locally more than half of the St-Victor synclinorium is masked.

Near St-Georges-de-Beauce and Lake St. Francis the folds are tight, isoclinal, gently plunging, and generally overturned to the southeast. Southeast of Sherbrooke, the St. Francis is almost everywhere overturned to the southeast and the tops of all beds face southeast. This extraordinary homocline which represents the northwest limb of a major syncline, is maintained at least from Cookshire to Malvina.

REFERENCES

ALBEE, A. L. (1957). Bedrock geology of the Hyde Park quadrangle, Vermont. U.S. Geol. Surv., Geol. Quadrangle Map GQ102.

CHIDESTER, A. H. (1953). Geology of the talc deposits, Sterling Pond area, Stowe, Vermont. U.S. Geol. Surv., Min. Investigation Field Studies, Map M.F. 11.

DE RÖMER, H. (1960). Geology of the Eastman-Orford Lake Area, Eastern Townships, Province of Quebec. Ph.D. thesis, McGill University.

EAKINS, P. R. (1963). Sutton map-area, Quebec. Geol. Surv. Can., paper 63–34, 3 pp. map.

ERIC, J. H. and DENNIS, J. G. (1958). Geology of the Concord-Waterford area, Vermont. Vt. Geol. Surv., Bull. 11.

OSBERG, P. H. (1952). The Green Mountain Anticlinorium in the vicinity of Rochester and East Middlebury, Vermont. Vt. Geol. Surv., Bull. 5.

OSBERG, P. H. (1965). Structural geology of the Knowlton-Richmond Area, Quebec. Geol. Soc. Amer., Bull. 76, (2): 223–250.

RICKARD, M. J. (1965). Taconic orogeny in the western Appalachians: experimental application of microtextural studies to isotopic dating. Geol. Soc. Amer. Bull. 76 (5): 523–536.

ROTH, H. (1965). A structural study of the Sutton Mountains, Quebec. Ph.D. thesis, McGill University.

CONTRIBUTIONS FROM SYSTEMATIC STUDIES OF MINOR STRUCTURES IN THE SOUTHERN QUÉBEC APPALACHIANS

J. Béland

ABSTRACT

The gross features of the regional structure and stratigraphy of the western part of the Eastern Townships of southern Quebec are outlined to supply a framework to a review of several studies of minor structural features (foliations, lineations, and minor folds) that have been conducted in this region. These studies indicate quite clearly that this segment of the Northern Appalachians has been submitted to several phases of deformation. The relations of these phases to the Taconic and Acadian orogenies can be inferred to some extent from unconformities and isotopic dating of metamorphic minerals.

The studies also show that as time elapsed the tectonic events changed in character. The early folding was apparently of the recumbent type variable in trend and may be related to some form of gravitational movement. The late phase, in contrast, produced upright folds, consistently oriented and very likely induced by compressive tangential stresses. Some late cleavages suggest cleavage domes possibly reulting from the diapiric rise of granitoid bodies late in the tectonic history of the region.

THE QUESTION that is here raised is that of the contribution which systematic studies of minor structures such as foliations, lineations, and minor folds have brought to the clarification of the tectonic history of the southern Quebec Appalachians. The immediate answer is that it has been a major contribution.

Such studies have been conducted in conjunction with aerial mapping by several workers. Figure 1 shows the regional geology with the locations of the areas covered. The first detailed structural study in the region was by Osberg (1965) who, between 1953 and 1959, mapped about 300 square miles south of the St-François River. De Römer (1961), while mapping the St-Etienne-de-Bolton area, in the southeastern part of the territory covered by Osberg, collected similar information. From 1957 to 1960 Rickard (1965) studied an area of about 400 square miles just south of the ground examined by Osberg. In 1960 and 1961 Baer (1961) carried the same type of study over an area of some 60 square miles located in the Stoke Range, southwest of Sherbrooke.

All these studies involved systematic measurements of attitudes of bedding, schistosities, cleavages, lineations, and minor fold axes, and Rickard's study included isotopic dating of metamorphic minerals formed along certain cleavage planes. Each study led to the reconstitution of a succession of tectonic events which have been compiled in Table I. The numbering used

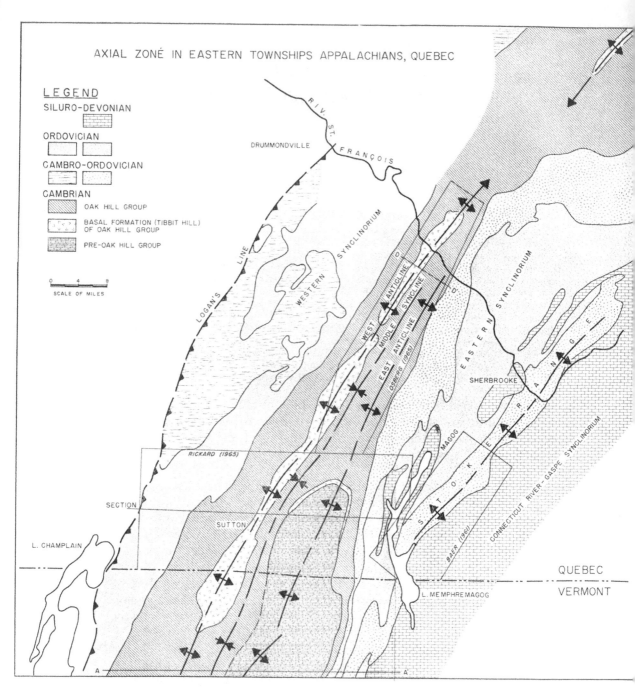

FIGURE 1

by the authors to indicate the chronological order of events has been modified in Table I so as to achieve uniformity and to relate the various successions proposed. But first let us consider the regional setting in which these studies were made.

REGIONAL GEOLOGY

Figure 1 was intended to show the gross geological features of the region which covers the western part of the Eastern Townships of Quebec approximately from the St-François River to the Quebec-Vermont boundary between Logan's Line on the west and the western margin of the Connecticut River-Gaspé synclinorium on the east. Logan's Line on the west marks the boundary between the foreland and the Northern Appalachian folded belt.

The main feature in this segment of the Northern Appalachians is the Sutton-Green Mountains anticlinorium which here is a northeasterly plunging, double anticline along which mildly to moderately metamorphosed rocks crop out. Considering that this metamorphism dies out on either flank, the anticlinorium may be termed an axial zone. For simplicity of reference and to avoid the multiplicity of names that have been given to the two anticlines and the intervening syncline from both sides of the border they are here called East and West anticlines and Middle syncline (Figs. 1 and 2). Each of these structures is actually made of a number of folds.

Northeast of the St-François River the general northeasterly plunge of the anticlinorium is apparently reversed. As little is known of the detailed structural features of this area it will not be discussed here.

The metamorphic rocks of the axial zone are greenschists, mica-schists, occasionally garnetiferous, and low grade amphibolites of sedimentary and volcanic origins. Marble occurs in minor amounts. The bulk of the metasediments is believed to have been pelites and pelitic sandstones. In Figure 1 all these rocks are grouped into three units: (1) pre-Oak Hill Series, (2) Tibbit Hill Formation of the Oak Hill Series, and (3) the remainder of the Oak Hill Series. The Tibbit Hill Formation, which is the base of the Oak Hill Series as established by Clark (1934), was used as a marker because, being made almost exclusively of metavolcanic rocks, it is easily traced. The Oak Hill Series (or Group) has been dated from fossils as Cambrian and the underlying unit, which rests on Precambrian rocks farther south, is also thought to be Cambrian.

On either side, the Sutton-Green Mountains anticlinorium is bordered by synclinoria. The rocks of the western synclinorium, considered to be Cambrian and Ordovician, are here divided into two units. The most widespread includes pelites some of which, thought to belong to the foreland, have been referred to the St-Germain Complex (Eakins 1964). The rest of these pelites, which are presumably of geosynclinal affinity, have been grouped under the name Stanbridge Formation (formerly Farnham). The other unit, more restricted than the first one, is an assemblage of sandstone

TABLE I
Successive Folding in Cambro-Ordovician Rocks

Author	Folding		Cleavage	Age
Osberg (1965) West and East Anticlines	F_1 N 60 W isoclinal folding, slip movement	S_1	Flow, axial plane	Taconic?
	F_2 N 50 E isoclinal folding, slip movement	S_2	Flow, axial plane	Acadian?
	F_3 N 30 E open folding, flexure and slip movement	S_3	Flow and fracture, axial plane	Acadian?
De Römer (1961) East Anticline	F_1 E-W recumbent folding, slip movement	S_1	Flow, axial plane	Taconic
	F_2 N 15 E open folding, flexure and slip movement	S_2	Fracture and flow, axial plane	Taconic
Rickard (1965) West and East Anticlines and Eastern Synclinorium	F_0 Nappes			Middle Ordovician
	F_1 (West anticline) N-S upright folding (East anticline) E-W recumbent folding (Eastern synclinorium) N-S isoclinal folding, steep plunge Thrusting	S_1	Flow, axial plane, steep dip Flow, axial plane, flat Flow, axial plane, steep dip	Taconic
	F_2 (West anticline and Eastern synclinorium) no folding (East anticline) N-S open folding	S_2	Fracture, oblique Flow, axial plane	Taconic
Baer (1961) Stoke Range Anticline	F_1 NE macrofolds	S_1	Flow, axial plane	Acadian
	F_2 Arcs, minor folds by diapirism	S_2	Flow, oblique, steep dip	Acadian
	F_3 Arcs, minor folds by diapirism	S_3	Fracture, oblique, moderate dip	Acadian

and pelite (greywacke-shale) known as the Sillery Formation. Opinions vary as to what the relationships of the two units are. Some think that the Sillery is conformably above the Oak Hill Series and below the Stanbridge (Cady 1960, p. 558) whereas others are inclined to think that the Sillery is in fault contact with the adjacent pelites and may have been thrust from the east in nappe-like fashion over or into rocks that properly belong to the foreland.

The eastern synclinorium extends to the Stoke Range anticline and its rocks, probably all Ordovician, can be roughly divided into two units. The basal unit alongside the axial zone includes volcanic rocks, serpentinite intrusives, abundant coarse-grained to conglomeratic sandstones and some pelites. Conformably above this is the second unit which is mostly pelites. At the Stoke Range anticline the basal unit with all its components reappears.

At the eastern limit of the region here treated the Gaspé-Connecticut River synclinorium begins. It includes a basal conglomerate followed by limestone and limy pelites. Other rocks show up higher in the sequence farther east. Fossils point to Silurian and Devonian ages. Near Lake Memphré-magog and farther east stocks of granite and granodiorite intrude the Siluro-Devonian strata.

A severe deformation is observed in this region in both the Cambro-Ordovician and the Siluro-Devonian strata. As in other parts of the Northern Appalachians, folding could be ascribed to the Taconic and Acadian orogenies, respectively of Late Ordovician and Middle to Late Devonian age. The marked angular unconformity which in some other parts of the Northern Appalachians allows inferences of the two orogenies is not, however, easily seen in the Eastern Townships and the possibility that there has been here but one major mountain building period at the close of the Devonian sedimentation (Cady 1960) must be considered.

So let us see now what light the study of minor structural features and the dating of metamorphic minerals have thrown on this subject.

DETAILED STRUCTURAL ANALYSIS

(a) Osberg's study

The close examination and elaborate statistical treatment Osberg (1965) has made of the many S-planes, lineations, and minor folds he has encountered from the West to the East anticlines in the northern part of the axial zone, have led him to conclude that in this zone there have been probably three successive phases of folding (F_1, F_2, F_3 in table I). The predominant tectonic grain, about N30°E, is ascribed to the last episode, F_3. The preceding phases F_1 and F_2 produced folds that trended respectively N60°W and N50°E. He remarks also that the East anticline shows a more complex structural pattern than the West anticline. Although both structures have been subjected to at least two foldings, the West anticline shows simple open folds outlined by bedding and but slightly overturned to the northwest, whereas the form surface of the East anticline is not bedding but an earlier schistosity, S_1 or S_2, which was presumably an axial-plane schistosity. Osberg (personal communication) considers that perhaps different lithologies and consequently different competencies have led to different styles of deformation.

In view of the habitual chronology of orogenic events used in the Northern Appalachians, one might be tempted to assign F_1 and F_2 to the Taconic orogeny and F_3 to the Acadian episode since the trend of F_3 is the only one consistent with the gross alignment of the folds observed in the nearby Siluro-Devonian strata. Osberg, however, is of a different opinion. He considers F_2 and F_3 as probably Acadian because they match trends observed in some areas of Siluro-Devonian rocks in Vermont. F_1 is thought to be Taconic because metamorphic minerals collected from the axial zone farther south in Quebec have given Ordovician ages of 440 and 420 m.y.

Osberg also proposes structural sections across the axial zone which to the author of this paper convey the idea of tectonic transport away from the foreland during the early phases (F_1 or F_2) of folding, as for example section DD^1, reproduced in Figure 2. Osberg (personal communication), however, considers that the recumbency of the folds resulted from flattening produced by F_3 and that the direction of overturn is accidental. It is interesting to note, in relation to this, that some structural sections attached to the Vermont Centennial Geologic Map (Doll 1961), as for example section AA^1 (also reproduced in Fig. 2), indicate folding of the same nature but overturned towards the foreland. It seems therefore that either the sense of transport was not consistent throughout the region or that the direction of overturn in the East anticline has no significance. Sections DD^1 and AA^1 are about 25 miles apart (see Fig. 1).

(b) De Römer's study

De Römer (1961) made his observations in an area included in the territory that was covered by Osberg. His conclusions are much in accord with those of Osberg except that only two phases of deformation (F_1 and F_2 in Table I) are inferred. The early phase, F_1, is thought to have led to the formation of isoclinal recumbent folds that trended about east-west and were later arched along a N15°E axis so as to produce the East anticline. Both phases are considered to be probably Taconic because of the marked angular unconformity believed to exist between the Ordovician strata and the Siluro-Devonian cover resting on them at Lake Memphrémagog. In a recent communication to the author, De Römer indicates that a re-study of the area has led him to the conclusion that a still earlier phase preceded the two phases mentioned in this paper of 1961. This earlier phase is said to be revealed by a cleavage refolded by F_1. Thus, as was found by Osberg, three phases of deformation can be here reconstructed.

(c) Rickard's study

Rickard (1965), who included in his study, besides the axial zone, parts of the bordering synclinoria (Fig. 1), agrees with Osberg's three phases of

STRUCTURAL   SECTIONS
EASTERN   TOWNSHIPS   AND   VERMONT   APPALACHIANS

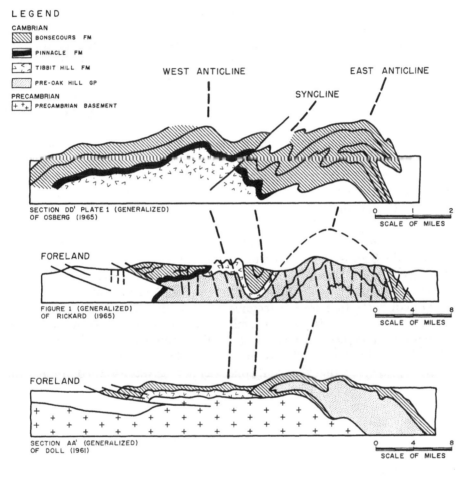

FIGURE 2

folding (F_0, F_1 and F_2 in Table I). The first phase, F_0 consisted, however, in the gliding of nappes descending from the east and is not recorded in the minor structures. It is for this reason labeled F_0 in Table I. The role given to phases F_1 and F_2 is much in accord with what was proposed by De Römer (1961) and, with some modification, consistent also with the picture outlined by Osberg. The early phase, F_1, is said to have produced, in the West anticline, folds gradually overturned towards the west as the foreland is approached (see middle section of Fig. 2); however, in the East anticline, the early folding took the form of isoclinal recumbent folds with approximately east-west axes and during the late phase F_2 the folds were refolded into the broad arch of the East anticline. Thus the form surface of the East anticline is given by the early axial-plane schistosity (or cleavage) S_1. Because of this arching early east-west folds have a very steep plunge on the flank of the arch. These early, steeply plunging, isoclinal folds are said to persist to the the east in the western flank of the eastern synclinorium where they are truncated by the gently plunging tight synclines of Siluro-Devonian strata.

Rickard considers that the late folding F_2 was restricted to the East anticline and east and west of that structure it is replaced by fracture cleavage becoming fainter as the foreland or the eastern synclinorium are approached. The East anticline is thought to have been produced by slip movement induced by lateral compression as evidenced by minor asymmetrical folds open towards the crest of the arch.

As to the chronology of events, Rickard holds that phase F_0 probably took place in Middle Ordovician time and that phases F_1 and F_2 are Taconic. F_1 and F_2 are believed to be Taconic because completely recrystallized micas, oriented along the S_2 planes at the East anticline, have yielded an Ordovician age of 440 m.y. therefore excluding any possibility of a major reworking of these rocks in post-Ordovician time.

This Taconic dating of the late cleavage, S_2, is somehow unexpected because the Siluro-Devonian strata on the west side of Lake Memphremagog, only a short distance away from the axial zone, are tightly folded and highly cleaved. It was because of this severe deformation in the Siluro-Devonian rocks that on the tectonic map of the Canadian Appalachians (Neale *et al.* 1961) the western limit of the Acadian folding in the Eastern Townships had been drawn west of the Sutton-Green Mountain anticlinorium. The isotopic dating of the late cleavage at 440 m.y. tends to indicate, however, that the axial zone was not, as was assumed, appreciably involved in the Acadian orogeny.

(d) Baer's study

The work of Baer (1961) in the Stoke Range anticline (Fig. 1) adds to the previous studies a new element by tentatively ascribing part of the cleavages and minor folds observed in that anticline to diapiric movements related to the emplacement of Acadian granitoid bodies such as are observed in the Gaspé-Connecticut River synclinorium nearby. One body, noted by Baer, crops out at the international border east of Lake Memphrémagog.

In the Stoke Range anticline Baer sees the effects of four successive phases of deformation of which the last one, which is jointing, can be neglected here. All three previous phases (F_1, F_2, and F_3 in Table I) involve the development of minor folds and cleavages. The cleavages S_2 and S_3 which delineate at the surface a series of broad arcs and which both dip northwesterly, with S_2 steeper than S_3, are considered to outline cleavage domes as are found around diapirs. Thus F_2 and F_3 are tentatively interpreted as doming phases which came after a folding phase to which the minor folds and cleavage of F_1 are ascribed. The doming phases could possibly be related to the diapiric rising of granitoid bodies assumed to have taken place as tension replaced the compressive stresses that prevailed during phase F_1. Baer considers that all the minor structures he has encountered in the Stoke Range anticline can be explained by the Acadian episode of deformation without having to resort to any earlier orogeny. He claims, besides, that wherever he could observe Ordovician and Siluro-Devonian rocks in contact both displayed the same minor structures.

CONCLUSIONS

The systematic studies of minor structural features conducted in the Sutton-Green Mountain axial zone by Osberg (1965), Rickard (1965), and De Römer (1961) point out quite clearly that this zone was submitted to several successive phases of deformation. The dating obtained from metamorphic minerals formed during the last phase tends to indicate that the deformations have to be Taconic or earlier and that no major reworking of these rocks took place after the Taconic orogeny had ceased. Perhaps the deformation took place almost continuously until that part of the folded belt achieved a high degree of rigidity that made it refractory to any further deformation.

Early folding at the East anticline apparently consisted of isoclinal recumbent folds that had variable trends. They run obliquely to perpendicularly to the trend assumed during the late folding. The style of the early folding and the marked variation in trend combined with movements towards and away from the foreland might be considered as indications that this deformation was not caused by tangential stresses set up by simple compression of the geosynclinal area against the foreland but possibly by gravitational gliding induced by some sort of vertical movement within that part of the depositional trough. Doming, for instance, might account for the variation in trend.

The late folding, in contrast, is very consistent in trend, different in style, less severe, and more restricted than the early folding. It could very well, as has been suggested, have been caused by compressive stresses set about perpendicularly to the trough after some degree of consolidation had been achieved by the early folding.

The manner in which the gently plunging synclines of Siluro-Devonian strata rest on the vertically tilted isoclinal early folds as pictured by Rickard (1965) points to a major angular unconformity and implies a wearing down of the folded belt prior to the accumulation of the Siluro-Devonian strata. Later, the Acadian folding took place, throwing the Siluro-Devonian cover into tight, highly cleaved, upright folds but without apparently inducing any recrystallization in the consolidated Cambro-Ordovician substratum. The fact that the Acadian folds conform in direction with those of late Taconic folding might be considered an indication that a permanent tectonic grain had been established in the mountain belt at the end of the Taconic episode.

The last event, at least as far as cleavages and minor folds are concerned, could have been that suggested by Baer (1961): the development of cleavage domes around granitoid bodies rising through the Cambro-Ordovician substratum. The possibility that these diapirs might represent mobilized Grenville basement remains to be explored.

REFERENCES

BAER, A. (1961). Structural analysis of Appalachian structures in the Memphrémagog area. Seminars in Tectonics, McGill University, mss. 20–22.

CADY, W. M. (1960). Stratigraphy and geotectonic relationships in northern Vermont and southern Quebec. Geol. Soc. Am. Bull., *71*: 531–576.

CLARK, T. H. (1934). Structure and stratigraphy of southern Quebec. Geol. Soc. Am. Bull., *45*: 1–20.

DE RÖMER, H. S. (1961). Structural elements in southern Quebec, northwestern Appalachians, Canada. Geol. Rundschau, *51*: 267–280.

DOLL, G. G. (1964). *Editor*, Centennial geologic map of Vermont. Vt. Geol. Surv.

EAKINS, P. R. (1964). Sutton map-area, Quebec. Geol. Surv. Can., Paper, 63–34.

NEALE, E. R. W., BÉLAND J., POTTER, R. R. and POOLE, W. H. (1961). A preliminary tectonic map of the Canadian Appalachian region based on age of folding. Bull. Can. Inst. Min. and Met., *54*: 687–694.

OSBERG, P. H. (1965). Structural geology of the Knowlton-Richmond area, Quebec. Geol. Soc. Am. Bull., *76*: 233–250.

RICKARD, M. J. (1965). Taconic orogeny in the Western Appalachians: experimental application of microtextural studies to isotopic dating. Geol. Soc. Am. Bull., *76*: 523–536.

GEOSYNCLINAL SETTING OF THE APPALACHIAN MOUNTAINS IN SOUTHEASTERN QUEBEC AND NORTHWESTERN NEW ENGLAND*

Wallace M. Cady

ABSTRACT

The Lower and Middle Paleozoic orthogeosyncline of the northern Appalachian Mountains includes a broad eugeosynclinal zone, a miogeosynclinal zone to the northwest and probably also to the southeast of the eugeosynclinal zone, and several geanticlines. The orthogeosyncline is adjoined on the northwest by the Precambrian basement rocks of the North American craton. Pelitic and semipelitic rocks dominate in the upper eugeosynclinal deposits and lap over the geanticlines, quasi-cratonic belts, and the margin of the craton. In the lowest Paleozoic rocks, the belt of transition between the northwestern miogeosynclinal zone and the eugeosynclinal zone lies northwest of the present Sutton-Green Mountain anticlinorium; it is missing in Quebec where the eugeosyclinal zone extends to the northwestern margin of the orthogeosyncline. But the belt of transition is farther and farther southeast in successively younger rocks, and in the Middle Paleozoic sections it approaches the northwestern boundaries of Maine and New Hampshire. The geanticlines, long structural highs including tectonic islands, are recognized by stratigraphic convergence and unconformable overlap toward their axes, and some are coincident with long gravity highs. The northwestern part of the orthogeosyncline and adjoining craton in Quebec are overlain by secondary geosynclinal deposits—an Upper Ordovician exogeosyncline. Early folds, similar in style, and related slides are features of the orthogeosyncline. Late folds, parallel in style, and related thrust faults are features of post-geosynclinal Middle Paleozoic anticlinoria and synclinoria that rudely coincide with the previously formed geanticlines and geosynclinal troughs.

THE NORTHWESTERN HALF of the northeastern extension of the Appalachian Mountain mobile belt includes the southeastern part of Quebec and adjoining portions of the New England States (Fig. 1). The geosynclinal features of the area are related to post-geosynclinal features, notably anticlinoria and synclinoria (Fig. 1) that are principally responsible for the pattern of geologic maps. The major geosynclinal features are shown on the paleotectonic map (Fig. 2) and in the restored sections (Fig. 3). The restored sections also diagram the distribution of unconformities, the directions in which early folds are overturned, and the intrusive relations of plutons.

The northwestern border of the northern part of the Lower and Middle Paleozoic Appalachian orthogeosyncline (Fig. 2) is adjoined on the northwest by the North American craton. Precambrian basement rocks crop out

*Publication authorized by the Director, the United States Geological Survey.

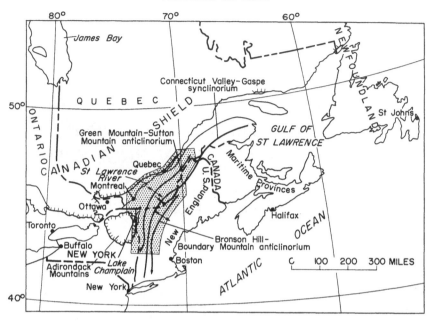

FIGURE 1. Index map showing location and major tectonic features of the region
of southeastern Quebec and northwestern New England.

in the Adirondack Mountains and Canadian Shield (Fig. 1), but are
covered by thin Paleozoic strata in the intervening Ottawa Valley embay-
ment. The orthogeosyncline includes a broad eugeosynclinal zone, a miogeosynclinal zone to the northwest next to the craton, and several geanticlines
(Cady 1960, p. 556). The presence of another miogeosynclinal zone, probably near the southeastern border of the orthogeosyncline, is suggested by
rocks transitional to miogeosynclinal in type in southeastern New England
(Emerson 1917, pp. 24–31, 35–39).

EUGEOSYNCLINAL AND MIOGEOSYNCLINAL ZONES

Both the eugeosynclinal and miogeosynclinal zones contain variously
metamorphosed Lower and Middle Paleozoic marine-bedded rocks, which
unconformably overlie the Precambrian basement. Those of the eugeosynclinal zone include slate, phyllite, schist, quartzite, granulite, partly foliate
granitic rocks, calc-silicate rocks, gneiss, greenstone, amphibolite, and
metarhyolite. Their protoliths include interbedded shale and graywacke
(collectively semipelites), calcareous siltstone, and mafic to felsic volcanic
rocks. Those of the miogeosynclinal zone include marble, quartzite, and
slate, the protoliths of which are fairly obviously limestone or dolomite,
quartz sandstone, siltstone, and shale that near the cratonal margin are
unmetamorphosed. Pelitic and semipelitic rocks, especially abundant in the
upper part of the eugeosynclinal zone, overlap both the margin of the craton
and geanticlinal and quasi-cratonic tracts within the orthogeosyncline, and

these incompetent rocks form the cores of recumbent folds and the soles of slides.

The bedded rocks in the eugeosynclinal zone are from two to at least five times as thick as those in the miogeosynclinal zone (Fig. 3). The maximum thickness of the Lower Paleozoic rocks in the miogeosynclinal zone is about 10,000 feet, whereas the thickest sections of rocks of the same age in the eugeosynclinal zone are probably at least 50,000 feet. Mafic sediments were probably derived from volcanic islands. Possibly some, but certainly not all, of the felsic sediments were derived from geanticlinal or cratonal sources to the southeast in the region now covered by the Atlantic Ocean (see Naylor and Boucot 1965). A large source, including that of the overlapping pelites and semipelites, was probably in a northeast-trending belt of uplifted Lower Paleozoic eugeosynclinal rocks in southeastern New England. Although the presence of such an uplift in the geosynclinal rocks is unsupported to date by stratigraphic or structural evidence supplied by convergences of units and by unconformities, the belt appears to include the parting line between separate terranes of recumbent folds facing southeast and northwest.

The transition between the laterally coalescing miogeosynclinal and eugeosynclinal zones (Cady 1960, p. 557) is marked by contrasts in sedimentary facies (Logan 1862, pp. 323–325), total thickness of the contained rocks, and, where geanticlines intervene, thinning of stratigraphic sections (Fig. 3). This thinning is chiefly by convergence of isochrons, less commonly by unconformable overlap, toward the transition zone. The transition in lowest Paleozoic rocks lies in the St. Lawrence and Champlain valleys (Cady 1960, pl. 2; Doll and others 1961). In Quebec the eugeosynclinal zone apparently extends northwestward to the northwestern margin of the orthogeosyncline, and no miogeosynclinal zone intervenes (Cady 1960, pp. 557–558). This is indicated chiefly by the occurrence of Cambrian and Lower Ordovician rocks characteristic of the eugeosynclinal zone—graywacke, shale, and mafic to intermediate volcanic rocks—close to the Precambrian basement rocks of the craton and containing clasts derived from the adjacent cratonic basement (Osborne 1956, pp. 172–191, 197). The scarcity there of characteristic miogeosynclinal rocks—quartz sandstone, limestone, and dolomite—has been attributed to overthrusting of the eugeosynclinal rocks upon them; but evidence for the extensive thrusting implied is equivocal (Clark 1964c, pp. 75–80; Osborne 1956, pp. 170–173), and it is difficult to explain how the coarse clasts, derived from the craton, crossed such a hypothetical miogeosynclinal zone.

The transition from the miogeosynclinal zone to the eugeosynclinal zone is found farther and farther southeast in successively younger rocks (Fig. 2, 3), so that in the Middle Paleozoic the southeastern edge of the miogeosynclinal zone approaches, and to the northeast oversteps, the northwestern boundaries of Maine and New Hampshire (E. L. Boudette, personal communication, 1966; Cady 1960, p. 557, pl. 2; Marleau 1958, p. 109; 1959,

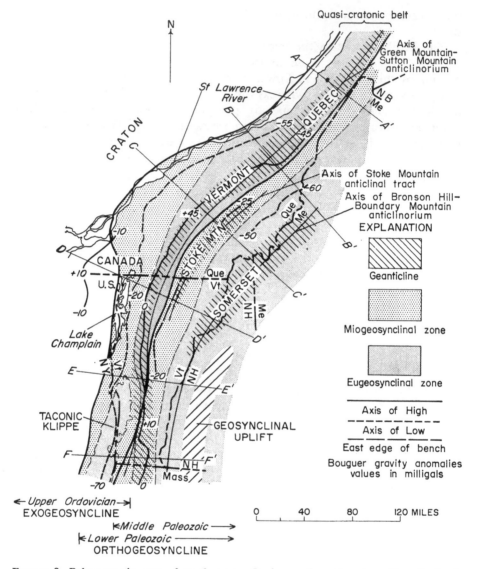

FIGURE 2. Paleotectonic map of southeastern Quebec and northwestern New England.

p. 137). Comparatively thin quartzose and calcareous units, whose affinities are miogeosynclinal although they are interbedded in the eugeosynclinal zone, extend southeast into these states (Cady 1960, pp. 561–562). This southeastward overlap of the miogeosynclinal zone on the eugeosynclinal zone implies stabilization of the northwestern part of the orthogeosyncline and probable conversion of the Lower Paleozoic miogeosynclinal zone to a quasi-cratonic belt (Fig. 2) (Cady 1960, pl. 2, p. 562). Concomitant southeastward offlap of the eugeosynclinal zone is shown by the distribution, both horizontally and vertically, of mafic and intermediate (?) metavolcanic and intrusive rocks and of ultramafic plutons (Cady and Chidester 1957).

Unconformities indicate stillstand mainly in the miogeosynclinal zone, and general uplift followed by subaerial denudation of geanticlines in both the eugeosynclinal and miogeosynclinal zones. Within the Lower Paleozoic sections, and chiefly beneath Middle Ordovician strata, unconformities (Fig. 3), overlaps, and stratigraphic convergences mark movements that

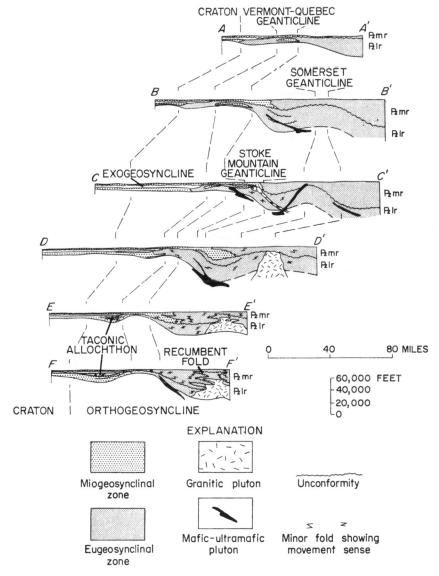

FIGURE 3. Restored sections of the lower (Pzlr) and Middle (Pzmr) Palaeozoic of southeastern Quebec and northwestern New England.

were forerunners of the Late Ordovician and Early Silurian Taconic disturbance (Albee 1957; Cady 1945, pp. 537–539, 560; 1960, p. 564; Cady, Albee, and Chidester 1963, p. 26; Lamarche, 1965, pp. 60–61, 144–150; Riordon, 1954, p. 8; 1957; St-Julien 1963a, pp. 30, 32, 125, 143–144, 193–195; Zen 1964, pp. 28–30). At the base of the Middle Paleozoic section in the miogeosynclinal zone, where the latter laps southeastward over the eugeosynclinal zone, is the most extensive unconformity within the orthogeosynclinal rocks; it marks the close of general uplift and deformation of the orthogeosynclinal belt, in the Taconic disturbance. Although it records a time interval sufficiently long to allow for erosional unroofing of Lower Paleozoic plutons, the unconformity and the plutons signify only temporary lapse of the eugeosynclinal regime, as shown by the reappearance of the mafic volcanic rocks in the Middle Paleozoic southeast of the belt of the overlap (Cady 1960, pp. 563, 565).

GEANTICLINES

Several geanticlines, long structural highs in the orthogeosyncline including paleotectonic islands, parallel the geosynclinal trend in northwestern New England and adjacent Quebec (Fig. 2). Convergence and unconformable overlap of bedded rock units toward their axes show their existence (Fig. 3). The geanticlines appear to have been sites of slowed deposition, episodic nondeposition, and local erosion during general uplift of the orthogeosynclinal belt. Two of the three known geanticlines coincide with gravity highs that apparently are produced by the high level of dense basement rocks (Fig. 2).

(a) Vermont-Quebec geanticline

The Lower Paleozoic (Cambrian and Ordovician) Vermont-Quebec geanticline[1] is northwesternmost. It is northwest of the axis of the younger Green Mountain-Sutton Mountain anticlinorium in northwestern Vermont and neighboring parts of Quebec, but swings into rude coincidence with this axis in west-central and southern Vermont, and to the northeast in Quebec (Fig. 2). It trends north along the Lower Paleozoic belt of transition from the miogeosynclinal zone to the eugeosynclinal zone in New England and nearby Quebec, but farther north its northeastward trend carries it within the eugeosynclinal zone. Its distribution is very nearly the same as that of a gravity high (Fig. 2) (Diment 1953; Fitzpatrick 1960).

The Vermont-Quebec geanticline is marked in northwestern Vermont

[1]The designation "Vermont-Quebec geanticline" takes the place of "Quebec Barrier," originally proposed (Ulrich and Schuchert 1902, p. 639) as a continuous linear subaerial topographic feature believed to have separated the belt in which rocks of the eugeosynclinal zone were deposited from that of the rocks of the miogeosynclinal zone. More recent studies have indicated that such a complete separation of eugeosynclinal and miogeosynclinal zones does not exist in Vermont and Quebec (Cady 1945, p. 561; 1960, p. 557) and probably is very rare in other regions (Kay 1951, p. 67).

and adjacent Quebec by unconformities in the eastern part of the miogeo-synclinical zone (Cady 1945, pp. 537–538; 1960, pp. 538–564), and by overlap of Middle Ordovician rocks upon Upper Cambrian ones (Cady 1960, pp. 541, 549, pl. 2) near the transition into the eugeosynclinal zone. The stratigraphic section thins toward the axis of the geanticline. The absence of 3,000 feet of Lower Ordovician strata in northwestern Vermont reduces the miogeosynclinal section on the west flank of the geanticline to about 7,000 feet, as compared with the 10,000 feet of Cambrian and Lower and Middle Ordovician strata found to the west near northern Lake Champlain.

To the north, in Quebec, presence of the Vermont-Quebec geanticline is shown by as much as tenfold stratigraphic convergence of Cambrian bedded rocks (Cady 1960, p. 538; Clark 1934, p. 10, table 2), toward the geanticlinal axis. The geanticline is also shown, where it trends northeast-ward from the miogeosynclinal into the eugeosynclinal zone (Fig. 3), by the intermittent occurrence of quartz sandstone, dolomite, and limestone units in the Cambrian (Cady 1960, pp. 539, 558). These features imply stillstand relative to adjacent subsiding parts of the eugeosynclinal zone.

Lower Paleozoic rocks have all been eroded in west-central Vermont, but indirect evidence indicates that the trend of the Vermont-Quebec geanticline there coincides with the younger Green Mountain-Sutton Mountain anticlinorium (Fig. 2). Middle Ordovician strata, west of the anticlinorium, overlap unconformably eastward on west-dipping and pro-gressively older Ordovician and Cambrian units apparently on the west flank of the geanticline (Cady 1945, p. 560; Fowler 1950, pp. 35–37; Zen 1964, pl. 2, sections B-B' and C-C').

The position of the geanticline in west-central Vermont is also suggested by the Lower Paleozoic section in the northern part of the Taconic klippe (Fig. 2), remnant of an allochthon (Fig. 3) that moved westward from the vicinity of the later site of the Green Mountain-Sutton Mountain anticlino-rium (Hawkes 1941, pp. 661–663). The Cambrian and Ordovician section in the Taconic klippe is thinner throughout than it is in either the eugeo-synclinal zone east of the anticlinorium or in the autochthonous miogeo-synclinal section to the west, beneath the klippe (Osberg 1959, pp. 45–46; Theokritoff 1964; Thompson 1952, p. 40; Zen 1961, p. 297, 331–333). The rocks of the klippe are lithically and sequentially most like the Lower Paleozoic eugeosynclinal rocks exposed down the plunge of the anticlinorium in northern Vermont and adjacent Quebec (Doll and others 1963, p. 95; Zen 1961, p. 333). The section in the Taconic klippe is there-fore believed to have been deposited atop the Vermont-Quebec geanticline. The section's relative thinness is thus explained by stratigraphic convergence and unconformable overlap of adjoining eugeosynclinal rocks to the east and miogeosynclinal rocks to the west, toward the geanticline. Tectonic thinning does not appear to have been involved, inasmuch as even some of the least deformed sections are very thin (Theokritoff 1964). The Vermont-

Quebec geanticline therefore appears to have been a site of tectonic denudation from which the now rootless rocks of the Taconic klippe slid westward. The westward sliding suggests that some uplift of the eugeosynclinal accumulations to the east occurred, as well as subsidence of the miogeosynclinal zone to the west.

(b) Stoke Mountain geanticline

The Lower Paleozoic (Lower and Middle Ordovician) Stoke Mountain geanticline coincides with the younger Stoke Mountain anticlinal tract on the southeast limb of the Green Mountain-Sutton Mountain anticlinorium, mainly in southern Quebec (Fig. 2). It contains eugeosynclinal Lower Paleozoic rocks and is overlain by miogeosynclinal Middle Paleozoic ones, inasmuch as the Middle Paleozoic belt of transition southeastward from the miogeosynclinal to the eugeosynclinal zone is a little southeast. Upper Silurian and Lower Devonian miogeosynclinal rocks lap southeastward over the geanticline into the adjacent geosyncline. Several other relations that are connected with early beginnings of the consolidation of the orthogeosyncline with the craton (Cady 1960, pp. 556–557) characterize the Stoke Mountain geanticline: (1) A Middle Ordovician epieugeosyncline is proposed northwest of the Stoke Mountain geanticline in Quebec (Lamarche 1965, p. 151; St-Julien 1963a, pp. 183–184), where the section appears to be nonvolcanic, although the sedimentary contents compare with those of the rocks interbedded with volcanic rocks in the eugeosynclinal zone to the south in Vermont. (2) In the geanticlinal tract itself are exposed plutons (Fig. 3, section C-C') of Lower Ordovician(?) albite granite and albite rhyolite (Duquette 1960, map 1344; St-Julien 1963b, map 1466), the former of which supplied phenoclasts to Middle Ordovician basal conglomerate contained in this epieugeosyncline (Lamarche 1962; St-Julien 1963a, p. 125). (3) Lower or Middle Ordovician basal conglomerate in northern Vermont (Albee 1957; Cady, Albee, and Chidester 1963, pp. 25–27; Lamarche 1965, pp. 144–145) probably reflects the proximity of the southern extension of the Stoke Mountain geanticline. (4) The geanticline coincides with a poorly defined narrow belt of slightly higher gravity (Fig. 2) within a wider belt of southeastward-decreasing gravity southeast of the Vermont-Quebec geanticline (Diment 1953).

(c) Somerset geanticline

The Lower and Middle Paleozoic Somerset geanticline[2] trends northeast near the northwestern boundaries of Maine and New Hampshire and about coincident with the northern part of the Bronson Hill-Boundary Mountain anticlinorium (Fig. 2). Showing at its core are bedded rocks of the eugeosynclinal zone that have been variously assigned to the Cambrian and

[2]Boucot and others (1964, p. 76) have identified a local source in Somerset County, western Maine, of material in Silurian and Lower Devonian sedimentary rocks which they refer to as paleogeographic "Somerset Island." The term "geanticline" seems more generally appropriate, to include areas in which broad and continuous subaerial extent is less evident.

(or) Ordovician (Albee 1961, pp. C51–C52; Billings 1956, p. 98), to a "basement complex," possibly Precambrian (Boucot 1961, pp. 184–185), to the Middle Ordovician (Cady 1960, pp. 554, 561, pl. 3, annot. 21), and to the "pre-Silurian" (Marleau 1959, p. 133). Also showing in the geanticline are Upper Ordovician plutonic rocks (Highlandcroft Plutonic Series, Fig. 3, section *D-D'*) and its possible equivalents (Albee 1961, p. C51, fig. 168.1; Billings 1955; 1956, pp. 46–48, 121–122, 148; Doll and others 1961). If gravity highs are connected with this geanticline they are largely masked by gravity lows produced by later-formed low-density metamorphic and plutonic granitic rocks (see Joyner 1963, pl. 1).

The unconformity that marks the Taconic disturbance truncates the Somerset geanticline and plutons exposed in the geanticlinal tract. The geanticline is overlain successively by quartz conglomerate, quartz sandstone, and limestone of the Lower and Middle Silurian, especially in its central and southern portions (see also Albee 1961, pp. (C53–C54). This succession bespeaks stillstand and progressively lowered relief of the geanticline (Cady 1960, p. 562), approaching, although not fully achieving, the relatively stable condition of the miogeosynclinal zone. Elsewhere the Somerset geanticline appears to be overlapped unconformably by Silurian and Lower Devonian eugeosynclinal rocks. A section of 50,000 + feet of Ordovician to Devonian eugeosynclinal rocks contained in and overlying the geanticline (Billings, 1956, p. 7, 9; Green 1964, p. 11) may be compared with a section of possibly 65,000 feet of correlative eugeosynclinal and miogeosynclinal rocks west of the geanticline in north-central Vermont, and with eugeosynclinal rocks of unknown thickness southeast of the geanticline in New Hampshire and Maine (Fig. 3).

Exogeosyncline

Superimposed on the orthogeosyncline northwest of the Vermont-Quebec geanticline and on adjacent parts of the craton is an exogeosyncline (Figs. 2, 3) (Kay 1951, p. 17–20). This is a secondary geosyncline that contains uppermost Lower Paleozoic (Upper Ordovician) sandstone, shale, and limestone preserved in southeastern Quebec, where they overlie shales at the top of the Cambrian and Lower and Middle Ordovician miogeosynclinal zone and extend onto the craton (Clark 1947, pp. 6–15; 1955, pp. 19–36; 1964a, pp. 5–19; 1964b, pp. 32–41; 1964c, pp. 8–62). The southeastern extent of the exogeosyncline is indeterminate, inasmuch as the contained rocks are eroded from all but the St. Lawrence Valley area. However, the Vermont-Quebec geanticline, which supplied the clastic sediments, is the likely southeastern limit.

Early Folds and Slides

The early folds in the orthogeosynclinal belt are chiefly similar folds in which the beds are thinner on the limbs than at the axes. They involve only the Lower and Middle Paleozoic bedded rocks, and not the Precambrian basement. Slides, major recumbent folds, intrastratal intrusions of bedded rocks, syntectonic bodies of ultramafic rock, and, less commonly, folds with nearly vertical axial surfaces are among the early structural features. Notwithstanding their large size, the slides and recumbent folds are not readily recognized. This is because only in limited areas do they obviously affect the pattern of regional geologic maps (scale 1:250,000), and sedimentary structures indicating tops of beds are masked by metamorphism; moreover, they were re-deformed in late episodes of doming, arching, and regional folding. The axial surfaces of the early folds are raised in domes and arches and flexed in anticlinoria and synclinoria that follow the northeast trend of the orthogeosyncline and are principally responsible for the map pattern.

Most of the early folds are longitudinal folds overturned to the west-northwest (Fig. 3) toward the cratonal margin of the orthogeosynclinal belt. The exceptions are cross folds and related rodding that trend rudely northwest about at the right angles to the longitudinal structural trends of the region, and oblique folds that trend more northeast than do the longitudinal folds in Quebec (Osberg 1965, pp. 242–243) and north-northwest in Vermont (Osberg 1952, pl. 2, p. 83), merging with the north-northeast-trending longitudinal folds in the intervening area. The Middle Ordovician Taconic slide is beneath the Taconic klippe (Fig. 2), which appears to contain the principal early longitudinal folds reported west of the Vermont-Quebec geanticline (Zen 1961, pp. 314–321; *see also* Doll and others 1961, 1963; Zen, 1963).

LATE FOLDS AND THRUST FAULTS

The late folds and thrust faults are post-geosynclinal Middle Paleozoic features, referred to the Acadian orogeny (Cady 1945, p. 580; 1960, p. 564). Unlike the early folds, they involve the Precambrian basement. The folds are grossly parallel (concentric) in style, and form the anticlinoria that very roughly coincide with the previously formed geanticlines, and the synclinoria (Fig. 1) that coincide with the intervening and adjoining geosynclines, in the orthogeosynclinal belt. The thrust faults are east-dipping, essentially foreland thrusts, formed chiefly in the Lower Paleozoic miogeosynclinal zone by breaking of anticlinal hinges of the late folds northwest and west of the axial anticlines of the Green Mountain-Sutton Mountain anticlinorium (Cady 1945, p. 577–579).

ACKNOWLEDGMENTS

This study is an outgrowth of regional geologic mapping and investigations of talc and asbestos deposits conducted in Vermont by the United States Geological Survey. The author is especially indebted to Arden L. Albee and Alfred H. Chidester, his associates in this project, for continuing discussion and criticism of interpretation. Comparable guidance has also been provided by J. R. Béland and F. F. Osborne and their colleagues of the Department of Natural Resources, Quebec. The manuscript was reviewed by Warren Hamilton and L. R. Page. Drafts of manuscripts on closely related topics have been helpfully criticized by M. P. Billings, R. H. Jahns, R. Y. Lamarche, P. H. Osberg, M. J. Rickard, J. R. Rosenfeld, Pierre St-Julien, and W. S. White. The author is grateful to J. B. Thompson, Jr. for many discussions of the regional geology and to E-an Zen for continuing correspondence concerning relations of the rocks in the Taconic Range. Numerous other geologists who have contributed unpublished information, suggestions, and criticism include F. W. Benoit, A. J. Boucot, T. H. Clark, G. W. Crosby, J. G. Dennis, W. R. Diment, H. R. Dixon, C. G. Doll, Gilles Duquette, P. R. Eakins, M. M. Fitzpatrick, J. C. Green, R. H. Konig, J. B. Hadley, N. L. Hatch, Claude Hubert, P. J. Lespérance, R. A. Marleau, R. H. Moench, R. S. Naylor, H. S. de Römer, D. T. Secor, G. L. Snyder, and B. G. Woodland.

REFERENCES

ALBEE, A. L. (1957). Bedrock geology of the Hyde Park quadrangle, Vermont. U.S. Geol. Surv. Geol. Quad. Map GQ–102.
——— (1961). Boundary Mountain anticlinorium, west-central Maine and northern New Hampshire, *in* Short papers in the geological and hydrological sciences: U.S. Geol. Surv. Prof. Paper 424–C: C51–C54.
BILLINGS, M. P. (1955). Geologic map of New Hampshire: U.S. Geol. Survey.
——— (1956). The geology of New Hampshire, Pt. II: Bedrock geology. New Hampshire Plan. Devel. Comm., 203 p.
BOUCOT, A. J. (1961). Stratigraphy of the Moose River synclinorium, Maine. U.S. Geol. Surv. Bull. 1111-E: 153–188.
BOUCOT, A. J., FIELD, M. T., FLETCHER, RAYMOND, FORBES, W. H., NAYLOR, R. S., and PAVLIDES, LOUIS (1964). Reconnaissance bedrock geology of the Presque Isle quadrangle, Maine. Maine Geol. Surv. Quad. Mapping Ser. 2: 123 p.
CADY, W. M. (1945). Stratigraphy and structure of west-central Vermont. Geol. Soc. America Bull., *56*: 515–587.
——— (1960). Stratigraphic and geotectonic relationships in northern Vermont and southern Quebec. Geol. Soc. Amer. Bull., *71*: 531–576.
CADY, W. M., ALBEE, A. L., and CHIDESTER, A. H. (1963). Bedrock geology and asbestos deposits of the upper Missisquoi Valley and vicinity, Vermont. U.S. Geol. Surv. Bull. 1122-B: 78 p.
CADY, W. M., and CHIDESTER, A. H. (1957). Magmatic relationships in northern Vermont and southern Quebec [abs.]. Geol. Soc. Am. Bull., *68*: 1705.
CLARK, T. H. (1934). Structure and stratigraphy of southern Quebec. Geol. Soc. Am. Bull., *45*: 1–20.
——— (1947). Summary report on the St. Lawrence Lowlands south of the St. Lawrence River. Quebec Dept. Mines, Geol. Surveys Br., Prelim. Rept. 204: 18 p., and Map 642.
——— (1955). St. Jean-Beloeil area. Quebec Dept. Mines, Geol. Surv. Br., Geol. Rept. 66: 83 p.
——— (1964a). Upton area, Bagot, Drummond, Richelieu, St. Hyacinthe, and Yamaska counties. Quebec Dept. Nat. Res., Geol. Explor. Service, G. R. 100: 37 p.
——— (1964b). St. Hyacinthe area (west half), Bagot, St. Hyacinthe, and Shefford counties. Quebec Dept. Nat. Res., Geol. Explor. Service, G. R. 101: 128 p.
——— (1964c). Yamaska-Aston area, Nicolet, Yamaska, Berthier, Richelieu, and Drummond counties. Quebec Dept. Nat. Res., Geol. Explor. Service, G. R. 102: 192 p.

DIMENT, W. H. (1953). Unpublished Ph.D. dissertation, Harvard University.

DOLL, C. G., CADY, W. M., THOMPSON, J. B., JR., and BILLINGS, M. P. (1961). Centennial geologic map of Vermont. Vt Geol. Surv.

———— (1963). Reply to Zen's discussion of the centennial geologic map of Vermont. Am. Jour. Sci., *261*: 94–96.

DUQUETTE, GILLES (1960). Preliminary report on Gould area, Wolfe and Compton electoral districts. Quebec Dept. Mines, Mineral Deposits Br., Prelim. R. 432: 10 p.

EMERSON, B. K. (1917). Geology of Massachusetts and Rhode Island. U.S. Geol. Surv. Bull. 597: 289 p.

FITZPATRICK, M. M. (1960). Unpublished Ph.D. dissertation, Harvard University.

FOWLER, PHILLIP (1950). Stratigraphy and structure of the Castleton area, Vermont. Vt Geol. Surv. Bull. 2: 83 p.

GREEN, J. C. (1964). Stratigraphy and structure of the Boundary Mountain anticlinorium in the Errol quadrangle, New Hampshire-Maine. Geol. Soc. Am. Spec. Paper, *77*: 78 p.

HAWKES, H. E., JR. (1941). Roots of the Taconic fault in west-central Vermont. Geol. Soc. Am. Bull., *52*: 649–666.

JOYNER, W. B. (1963). Gravity in north-central New England. Geol. Soc. Am. Bull., *74*: 831–856.

KAY, MARSHALL (1951). North American geosynclines. Geol. Soc. Am. Mem. 48, 143 p.

LAMARCHE, R. Y. (1962). Unpublished M.Sc. thesis, université Laval.

———— (1965). Unpublished D.Sc. thesis, Université Laval.

LOGAN, W. E. (1862). Considerations relating to the Quebec group, and the upper copper-bearing rocks of Lake Superior. Am. Jour. Sci., 2d ser., *33*: 320–327.

MARLEAU, R. A. (1958). Unpublished D.Sc. thesis, Université Laval.

———— (1959). Age relations in the Lake Megantic Range, southern Quebec. Geol. Assoc. Can. Proc., *11*: 129–139.

NAYLOR, R. S., and BOUCOT, A. J. (1965). Origin and distribution of rocks of Ludlow age (Late Silurian) in the northern Appalachians. Am. Jour. Sci., *263*: 153–169.

OSBERG, P. H. (1952). The Green Mountain anticlinorium in the vicinity of Rochester and East Middlebury, Vt. Vt Geol. Surv. Bull., *5*: 127 p.

———— (1959). The stratigraphy and structure of the Coxe Mountain area, Vermont, *in* 51st New England Intercollegiate Geol. Conf. Guidebook, Stratigraphy and structure of west-central Vermont and adjacent New York. Rutland, Vermont, p. 45–52.

———— (1965). Structural geology of the Knowlton-Richmond area, Quebec. Geol. Soc. Am. Bull., *76*: 223–250.

OSBORNE, F. F. (1956). Geology near Quebec City. Naturaliste Canadien, *83*: 157–223.

RIORDEN, P. H. (1954). Preliminary report on the Thetford Mines-Black Lake area, Frontenac, Megantic, and Wolfe counties. Quebec Dept. Mines, Mineral Deposits Br., Prelim. R. 295: 23 p.

———— (1957). Evidence of a pre-Taconic orogeny in southeastern Quebec. Geol. Soc. Am. Bull., *68*: 389–394.

ST-JULIEN, PIERRE (1963a). Unpublished D.Sc. thesis, Université Laval.

———— (1963b). Preliminary report on Saint Elie d'Orford area, Sherbrooke and Richmond counties. Quebec Dept. Nat. Res., Mineral Deposits Br., Prelim. R. 492: 14 p.

THEOKRITOFF, GEORGE (1964). Taconic stratigraphy in northern Washington county, New York. Geol. Soc. Am. Bull., *75*: 171–190.

THOMPSON, J. B., JR. (1952). Southern Vermont *in* Guidebook for field trips in New England, Geology of the Appalachian Highlands of east-central New York, southern Vermont, and southern New Hampshire: Field trip No. 1. Geol. Soc. Am.: 14–23, 38–41.

ULRICH, E. O., and SCHUCHERT, CHARLES (1902). Paleozoic seas and barriers in eastern North America. New York State Mus. Bull., *52*: 633–663.

ZEN, E-AN (1961). Stratigraphy and structure at the north end of the Taconic Range in west-central Vermont. Geol. Soc. Am. Bull., *72*: 293–338.

———— (1963). Age and classification of some Taconic stratigraphic units on the centennial geologic map of Vermont: A discussion. Am. Jour. Sci., *261*: 92–94.

———— (1964). Stratigraphy and structure of a portion of the Castleton quadrangle, Vermont. Vt Geol. Surv. Bull., *25*: 70 p.

GRAVITY MEASUREMENTS IN APPALACHIA AND THEIR STRUCTURAL IMPLICATIONS

M. J. S. Innes and A. Argun-Weston

ABSTRACT

The history of gravity investigations in the Appalachian regions of Canada is reviewed. The main features of the gravitational field are examined, using crustal models, in the light of the surface geology and tectonic history. A marked gravity high over the Sutton-Green Mountain anticlinorium, and an adjacent gravity low to the west or northwest cannot be explained entirely by the surface geology, and deep-seated sources at the base of the crust or in the upper mantle, are necessary. These major gravity trends which mark the boundary between Appalachia and the Shield are persistent and extend south to Alabama, and northeast to Newfoundland. Statistical analyses show that Bouguer and free-air anomalies are 30 milligals greater over the Applachians than over the adjoining Precambrian Shield. The Appalachian region is undercompensated by 20 milligals while the Shield is overcompensated by about 10 milligals. These gravity results are inconsistent with the crustal thicknesses determined seismically, but consistent with the higher seismic velocities indicating a denser crust underlying Appalachia.

IN REVIEWING the history of gravity work in Canada one finds that the first measurements were made in the high latitudes in conjunction with Sir William Parry's voyages in 1819–20 to discover an east-west route through the Arctic Seas. The records show that the first gravity work in southern Canada was carried out three-quarters of a century later, by Commandant Defforges of the Geographical Service of the French army. In 1893 he made pendulum observations in the basement of the MacDonald Physics Building at McGill University in Montreal. His observations have historical significance, not only because they were the earliest in Canada, but also because they were part of a series of measurements at ten sites across the continent, to provide one of the first tests of isostasy in North America.

Since these historic measurements, the progress that has been made with gravity investigations in the Appalachian region is illustrated in Figure 1. Early measurements by the Dominion Observatory were almost entirely made with pendulums for isostatic and geodetic studies. Later pendulum observations were made to provide control for regional gravimeter surveys. Thirty-five pendulum stations, shown by black circles in Figure 1, were established in the Appalachian region during the first half of this century. An isostatic study of the Gaspé region during this early period, by F. J. Alcock, F.R.S.C. and the late A. H. Miller, F.R.S.C. (Alcock and Miller 1932), using both gravity and geodetic measurements, is of interest in the light of recent results. On the basis of positive gravity anomalies and large

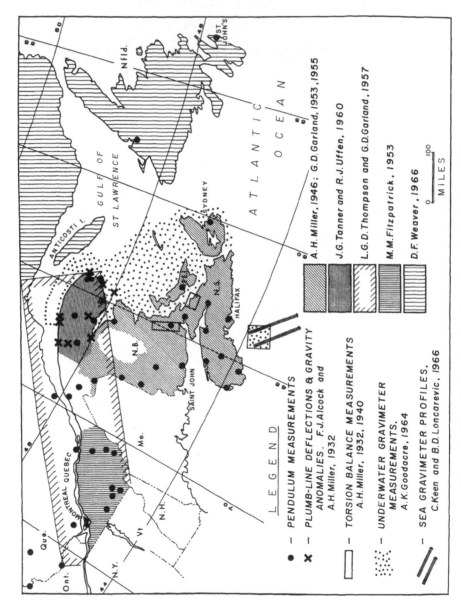

FIGURE 1. Map showing location of gravity work in the Appalachians.

deflections of the vertical at sites shown by crosses in Figure 1 these investigators deduced that the Gaspé region is locally undercompensated, that is, the mountains are without roots, and that a high-density crust underlies the waters of the Gulf of St. Lawrence to the east.

The Appalachian region also served as a proving ground, in the use of the torsion balance and magnetometer to solve problems of structural geology and of prospecting for minerals; for example A. H. Miller (1940) used these instruments to delineate buried ridges of pre-Carboniferous basement rocks near Moncton in eastern New Brunswick, and to outline the extent of the Malagash salt deposit in Nova Scotia.

Excellent progress has been made with regional gravity surveys since gravimeters were first used in 1946. With the exception of small areas west of the Matapédia River in Quebec, and parts of the highlands of New Brunswick and Nova Scotia, measurements at intervals of 12 km have been completed for most of the Appalachian region. Systematic underwater gravimeter measurements have been carried out by the Dominion Observatory in part of the Gulf of St. Lawrence, and surveys at sea using surface gravimeters have been initiated by the Bedford Institute of Oceanography. Nearly 14,000 regional stations now established within the Appalachian area are included in this review.

Only the major features of the gravitational field will be discussed briefly. Correlations between the gravity anomalies and surface geology will be pointed out and some speculations regarding the significance of other anomalies, showing no obvious correlation, will be given. More detailed discussion of the gravity features will be found elsewhere (Fitzpatrick 1953; Garland 1953, 1955; Thompson and Garland 1957; Tanner and Uffen 1960; Goodacre 1964a, 1964b; Goodacre and Nyland 1966; Weaver 1966a, 1966b, 1966c; Keen and Loncarevic 1966).

EASTERN TOWNSHIPS AND ADJOINING AREAS

For the purpose of discussion the area has been subdivided into four segments. The first area is that part of the province of Quebec south of the St. Lawrence River and west of the Chaudière River and includes the northern part of the states of New York, Vermont, and New Hampshire. As seen in Figure 2, the major structural divisions of this area are the Adirondack and Laurentian Highlands of the Precambrian Shield, the St. Lawrence Lowlands, and the western Appalachians. Figure 3 is a gravity map of the same area showing Bouguer gravity anomalies contoured at intervals of 10 milligals.

The anomalies are quite variable, ranging from −72 to +54 milligals for a total range of 126 milligals. The anomalies over the Shield areas are generally negative, compared to those over the Appalachian province. In the northern Adirondacks, the gravity field is highly variable, shows excellent

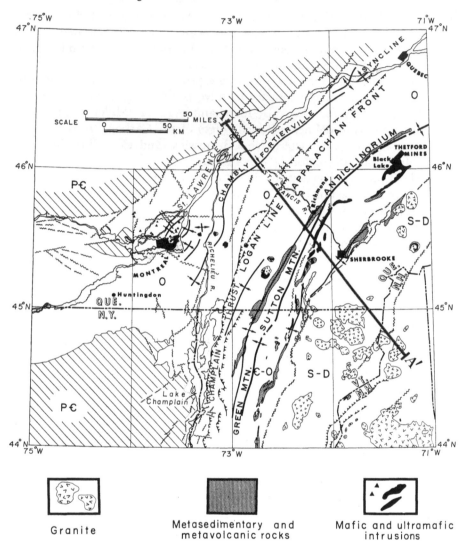

Granite

Metasedimentary and
metavolcanic rocks

Mafic and ultramafic
intrusions

FIGURE 2. Map showing major structural elements of south Quebec and the northern parts of the adjoining states of New York, Vermont, and New Hampshire.

correlation with the surface geology (Simmons 1964), and reaches minimum values of nearly −50 milligals over a large anorthosite body. In the Shield north of Montreal intensely negative anomalies which reach minimum values of −70 milligals or lower, correlate with low density granitic rocks. The most negative gravity values, as in the Adirondacks, have been found (Thompson and Garland 1957) to correlate with anorthosite, in this case the Morin anorthosite body. To the northeast (near Québec) the anomalies are more positive, and although these may be due in part to denser phases of the Grenville rocks, it is also possible that the anomalies

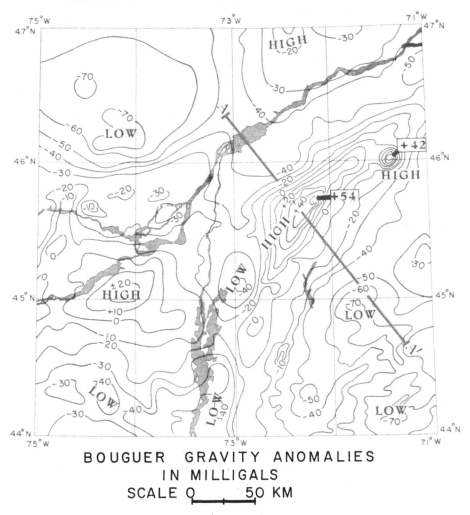

BOUGUER GRAVITY ANOMALIES
IN MILLIGALS
SCALE 0_____50 KM

FIGURE 3. Bouguer gravity anomaly map of the area shown in Figure 2.

reflect a deep-seated disturbance related to the deformation of the northern Appalachians.

The southeast portion of the map area is characterized by negative anomalies that correlate directly with exposures of granite intrusive rocks, of the White Mountain and New Hampshire plutonic series. Negative gravity anomalies are also prevalent to the northeast, forming a belt paralleling the Quebec-Maine boundary. Although the exposures of granite are not extensive and are few in number, it is suggested that these anomalies may reflect extensive granitic masses at depth.

South of the St. Lawrence River the anomalies trend northeast conforming to the structural trends of the Appalachians. A negative belt reaching a minimum value of −45 milligals parallels the river and corresponds in position to the canoe-shaped Chambly-Fortierville syncline (Clark 1956, 1961). Thompson and Garland (1957) have suggested that this gravity depression may be reasonably explained by the thicknesses of sedimentary strata determined by drilling. However, the mean density of the Ordovician and Cambrian rocks may not differ significantly from the density of the basement rocks, and another explanation for the negative gravity field may be necessary.

The most prominent gravitational feature of the map area is the large positive anomaly belt that coincides with the Sutton and Green mountains. While the relations are generally the same throughout this map area, for a distance of at least 320 km. the anomalies reach their maximum amplitude of +54 milligals, some 12 km. west of Richmond, Quebec. This is one of the most positive anomalies in Canada, rising more than 90 milligals above a background value of −40 milligals, in a distance of about 32 km. The steep gradients associated with this anomaly indicate that a considerable portion of the source lies high within the crust, the maximum depth to the top of the disturbing mass being less than 15 km.

It is clear that the gravity high closely marks the structural front of the northern Appalachians, separating the mildly folded miogeosynclinal rocks on the west from the highly deformed eugeosynclinal rocks to the east. This positive belt coincides with the uplifted metamorphosed Cambro-Ordovician strata of the Sutton-Green Mountain anticlinorium. Fitzpatrick's (1953) careful study, shows that the peak of the gravity anomaly closely follows the axis of the Tibbit Hill anticline, also referred to as the Enosburg Falls anticline (Cady 1960). Outcrops of basic and ultrabasic rocks appear to provide only minor control, as they do not coincide with the gravity maximum. An exception to this occurs in the Black Lake-Thetford area, where the gravity field reaches a local maximum of +42 milligals, reflecting both near surface ultrabasic rocks and the parent body at depth. These relations are illustrated better in Figure 4 which gives a generalized geological section and gravity profile across the Sutton-Green Mountain anticlinorium.

Undoubtedly the positive anomaly reflects, in part, heavy metavolcanic and metasedimentary rocks such as the Tibbit Hill schists, forming the core of the anticline, and also a possible thickening of the lower Cambro-Ordovician strata, as a result of overthrusting. However, some additional source is necessary; the integration of the total anomaly above regional background values indicates that if the Lower Palaeozoic are to be identified as the entire source, they must have the improbable thickness of 12 to 20 km. If the anomaly had originated entirely from a deep-seated source it would have required a major intrusion of mantle material into the crust. A wedge of mantle material 20 km wide at its base, rising to a height of 25 km. satisfies the gravity anomaly.

More convincing evidence that a deep-seated source is the probable

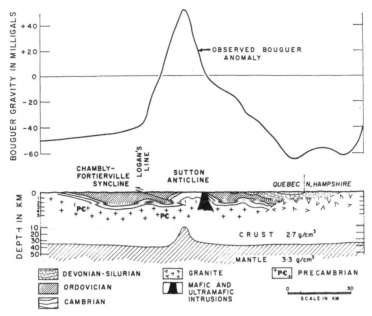

FIGURE 4. Generalized geological and crustal section and observed
Bouguer anomalies for profile A–A' (after Fitzpatrick 1953).

answer stems from gravity work in Vermont (Diment 1953). Figure 5 gives
an east-west generalized geological section and gravity profile across northern
New York and Vermont, near latitude 40°15′N. Rock sampling shows
variations in surface density along the section; but the variations do not

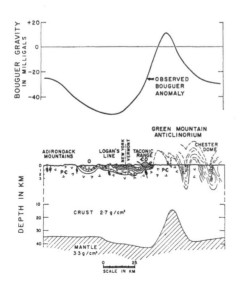

FIGURE 5. Generalized geological
and crustal section and Bouguer
anomaly profile across the Sutton-
Green Mountain anticlinorium
(after Diment 1953).

correlate with the gravity anomalies. For example, the peak anomaly occurs directly over the western exposure of uplifted Precambrian basement rocks of Grenville age, but the gravity field shows little or no correlation with similar rocks of the Chester Dome to the east. The Sutton-Green Mountain gravitational high also has been explained by a 20 km rise of mantle material into the crust. However, it should be pointed out that while deep-seated sources are required, they need not be abrupt changes at the base of the crust, as suggested, but could be a general increase in density throughout the whole crustal column. Such a condition could arise from large-scale penetration of basic material to higher levels during orogeny. The negative anomaly in the western part of the profile is part of an extensive negative belt which parallels the gravity high on the north and west to its termination in Alabama. This anomaly also cannot be explained entirely by the surface geology and a deep-seated source is indicated (Nettleton 1941; Woollard 1962).

Another striking gravitational feature in Figure 3 is a broad circular gravity high centred near Huntingdon, Quebec, in the St. Lawrence Lowlands some 40 miles southwest of Montreal. This positive anomaly, which reaches a high of $+25$ milligals is in an area underlain by Lower Ordovician dolomite and limestone. Independent studies of this anomaly by Sobczak (1966) and Simmons (1964) show that these strata, having a mean density of about 2.7 g/cm^3 and a thickness of less than 5,000 feet, would contribute very little to the gravity field. Although not dismissing the possibility that the origin lies in lithological changes in the Grenville rocks forming the basement, they postulate crustal warping with a 3 km rise of the crust-mantle boundary, as the main source of the anomaly. We might speculate further that such warping resulted in crustal rifting to form the severely block-faulted Ottawa-Bonnechère graben (Kay 1942). Positive anomalies have been noted in Africa and elsewhere, where crustal rifting and related alkaline intrusive activity have occurred. It is interesting to note that throughout the northern part of this predominantly positive area, numerous occurrences of alkaline intrusions, both large and small, are found (Clark 1952; Gold 1966). Examples of possible rifting in the Canadian shield are the Kapuskasing structure (Innes 1960) trending some 400 km south from James Bay and the Lake Timiskaming graben. Both structures are characterized by positive anomalies and situated on the axis of the former are six carbonatite alkaline ring complexes (McConnell 1966).

CHAUDIÈRE-MATAPÉDIA AREA

Figure 6 shows a section of the Appalachian region between the Chaudière River and the Matapédia valley. The available structural information (Dresser and Denis 1949; Béland 1961, 1962; Jones 1962) suggests that, compared to the area to the southwest, this region is simpler in structure; the folding has been less intense with only one major fold, an extension of

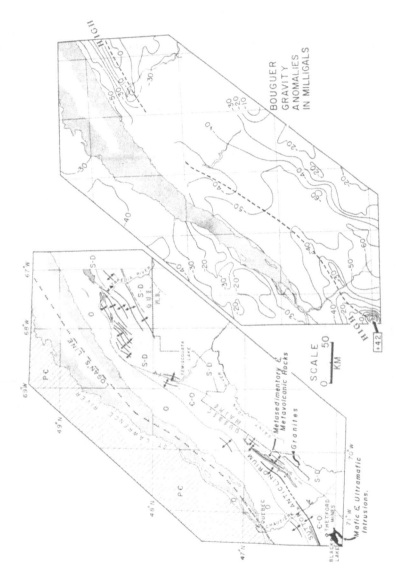

Figure 6. Major structural features of the Appalachian province south of the St. Lawrence River between the Chaudière and Matapédia Rivers and Bouguer gravity anomalies for the same area.

the Sutton-Green Mountain anticline, connecting the Eastern Townships with the Gaspé region. Correspondingly the gravity field is less disturbed. The most prominent gravitational feature is the negative anomaly belt paralleling the south shore of the St. Lawrence River where it reaches amplitudes as great as —60 milligals. The explanation for this negative belt is uncertain. It has been explained by Thompson and Garland (1957) as controlled partly by thickening of the Palaeozoic rocks due to thrusting against the shield, and partly by the Precambrian basement southeast from the St. Lawrence River being less dense. However, as will be seen later, the anomalies are generally more negative over the Precambrian areas than over the Appalachians, and may reflect a fundamental difference between the crustal columns of these geological provinces of contrasting age.

The extension of the positive belt over the Sutton-Green Mountain anticlinorium, may be identified as the positive anomalies that follow the exposure of Cambro-Ordovician rocks parallel to the Quebec-Maine boundary. Although not so intense as the former, this positive belt has been traced to the Gaspé except for the unsurveyed area in the vicinity of Lake Témiscouata.

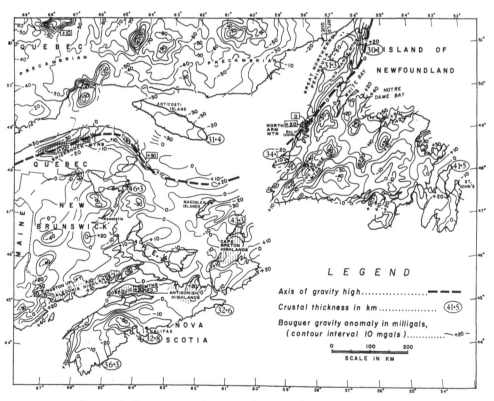

FIGURE 7. Bouguer gravity anomaly map of eastern Appalachia.

The Gaspé and the Maritime Provinces

The gravitational features of the Gaspé and Maritime Appalachians are shown in Figure 7. In the southern part of the map area Miller (1946) and Garland (1953) show that the major trends of the positive anomalies are related primarily to the structure of the pre-Carbiniferous basement rocks. In Nova Scotia and New Brunswick the positive anomalies correlate with the basement rocks forming the Highlands of Cape Breton, Antigonish, Caledonia, and the Cobequid Mountains. A major positive belt reflects the metasedimentary and metavolcanic rocks forming the core of the Kingston uplift. This structure, first investigated in 1940 near Moncton by A. H. Miller, with a torsion balance, has since been outlined in considerable detail by regional gravimeter work. It is at least 400 km long, strikes northeast across the straits of Northumberland and Prince Edward Island and terminates about 65 km west of the Magdalen Islands.

Negative gravity anomalies correlate with Devonian granites in southwestern Nova Scotia, and with those of the New Brunswick Highlands. It is possible that the circular negative anomaly centred about 50 km offshore from Newcastle, New Brunswick, also may be due to Devonian granite, as the thickness of sedimentary rock suggested by seismic work in the area is insufficient to account for the anomaly (Goodacre and Nyland 1966). Elsewhere the basins of Carboniferous and later sediments appear to produce negative anomalies. Analyses of the gravity field near the Magdalen Islands indicate the sedimentary strata to be about 3 km thick. The intricate pattern of the anomalies suggests a structure that may be highly folded and faulted.

The most outstanding positive anomaly belt occurs to the north in the Gaspé Peninsula. This east-west trending belt becomes most intense east of the Matapédia River, correlates with the Shickshock Mountains, and reaches a peak value of +35 milligals near the eastern end of the Gaspé Peninsula. Tanner and Uffen (1960) interpret these anomalies as due partly to high density metamorphic volcanic rocks of the Shickshock Series, that have been uplifted by faulting, and partly to dyke-like intrusions of ultrabasic rocks within the fault zone. Furthermore, a more complete explanation for the regional gravity field requires a rise in the intermediate layer, the existence of which is confirmed by seismic work in eastern Appalachia. Gravity data from underwater gravimeter surveys (Goodacre 1964) show that this structure trends some 300 km south and east from the Gaspé Peninsula to the limit of the survey directly north of the Magdalen Islands.

Newfoundland

The gravity field in Newfoundland is quite variable, the total anomaly range is about 100 milligals from a gravity low of −59 milligals over the Great Northern Peninsula, to +52 milligals in the Notre Dame Bay area. Weaver (1966a) in a comprehensive study shows that the anomalies are controlled largely by the surface geology. For example, the negative anomalies correlate with Devonian granite masses in southern Newfoundland, and to some extent, but not entirely, with Precambrian basement rocks forming the core of the Long Range Mountains of the Great Northern Peninsula. It is interesting to note that the Palaeozoic basement rocks within the Long Range Mountains in southern Newfoundland produce positive anomalies, as do the Palaeozoic basement rocks forming the highlands on the mainland. It is found that the Carboniferous sedimentary rocks in southwestern Newfoundland and those to the south of White Bay produce negative anomalies, but that other sedimentary rocks appear to have little effect on the gravity variations.

Most of the positive anomalies in central Newfoundland can be explained suitably by near-surface dioritic masses and/or gabbroic and ultrabasic intrusions. The residual positive anomalies indicate that near-surface dioritic masses extend to depths of 10 km or greater, while the positive regional field suggests that the crust for most of the island is abnormally dense.

Anomalies over gabbro and ultrabasic rocks in central Newfoundland show that these masses extend to similar depths, 7 to 10 km. On the other hand, the anomalies over the ultrabasic complex of the Bay of Islands indicate that these plutons are less deep, tending to support the theory that the Ordovician Humber Arm Group and basic intrusions are klippen that have been transported some 50 km from the east. A possible exception is the North Arm Mountain intrusion which produces a residual positive anomaly of 40 milligals corresponding to a thickness of 4 to 5 km. Weaver (1966a) suggests that such a thickness may be too great for gravity sliding, and that the North Arm Mountain ultrabasic body may have been intruded after movement of the clastic rocks, while other plutons further south, and those in the St. Anthony area, are part of the klippen zone.

One of the outstanding gravitational features of Newfoundland is the marked gravity gradient that trends northeasterly over the southern end of the Great Northern Peninsula. This line separates the predominantly negative anomalies of the Shield to the north from the predominantly positive anomalies of the Appalachians to the south. The anomalies appear to be caused by both near surface mass distributions and deep-seated disturbances within the crust or mantle. Weaver suggests that the positive anomalies of the Port au Port Peninsula in southwest Newfoundland are due in part to deep-seated basic intrusions and that this area may be structurally connected to the positive anomaly belt of the Gulf of St. Lawrence and the Gaspé region. Thus in a general way this whole anomaly belt, negative to the north and west and positive to the south, follows the Appalachian front from northern Newfoundland through Gaspé, the Eastern Townships, to the state of Vermont. Woollard's gravity map of the United States (1964) shows that this same gravity disturbance continues some 2,500 km south to Alabama.

Isostatic Considerations

According to Airy's classical "roots of mountains" theory, isostatic compensation of the earth's topography is accomplished by variations in crustal thickness. Analyses of measurements of gravity and elevation, coupled with seismic information have generally established the validity of this hypothesis in so far as the major topographic features of the earth are concerned. However an increasing number of seismic refraction studies, together with gravity data during the last few years are beginning to yield information to show that often, where there is a lack of isostatic equilibrium, the crust has both an anomalous thickness and an anomalous composition. Higher seismic velocities and positive anomalies appear to be characteristic of areas where the crust has a density and thickness greater than normal for its surface elevation. Similarly, in some areas of negative isostatic anomaly, seismic investigations find the crust to be abnormally thin. These conditions, first observed in restricted belts such as mid-oceanic ridges, island arcs, and rift valleys, have been found recently to exist in certain active orogenic belts. Although not completely understood, it is generally believed that such zones have resulted from an interchange of matter between the mantle and the crust (Cook 1962). Recent studies in the northern Alps (Cloos 1965) find a very considerable thickening of the basaltic layer, without a corresponding change in thickness of the granitic layer, suggesting that the thickening is the result of processes operating at depths between 20 and 70 km.

The geophysical results suggest that similar mass distributions underlie the northeastern part of the Appalachian province. Statistical analyses of the gravity data show that both the Appalachian area and the adjoining Precambrian Shield are not in isostatic equilibrium. The Appalachian region is undercompensated by about 20 milligals and the Shield overcompensated by about 10 milligals. Crustal thicknesses determined seismically by the Bedford Institute of Oceanography and Dalhousie University (Barrett et al. 1964; Ewing et al. 1966; Keen and Loncarevic 1966) and others, have been plotted on the map. In the central part of the Appalachians, which is characterized by positive anomalies, the crust varies in thickness from 41 to 45 km. An intermediate layer is present, and the velocities in the upper mantle are higher than average. On the other hand, areas of negative anomaly in the Shield, Anticosti Island, the Great Northern Peninsula, and granite areas of southeastern Nova Scotia have a thinner crust of 30 to 35 km, with no intermediate layer and with normal upper mantle velocities. There appears to be sufficient geophysical evidence to show that these major changes in crustal parameters take place along the boundary between the Precambrian Shield and the Appalachians. A full explanation for this phenomenon is a problem for the future.

References

Alcock, F. J., and Miller, A. H. (1932). Plumb-line deflections and gravity anomalies in Gaspé peninsula and their significance. Trans. Roy. Soc. Can., 26 (IV): 321–338.

Barrett, D. L., Berry, M., Blanchard, J. E., Keen, M. J., and McAllister, R. E. (1964). Seismic studies on the eastern seaboard of Canada: the Atlantic coast of Nova Scotia; Can. J. Earth Sci., 1: 10.

Béland, J. (1961). Some results of the application of the Russian concept of tectonic analysis in the Quebec Appalachians: Seminars in tectonics; Dept. of Geol. Sci., McGill Univ., 1, 1961.

——— (1962). Geology and petroleum possibilities of the Rimouski-Matapédia region; CIMM Bull., 55 (599): 158–161.

Cady, W. M. (1960). Stratigraphic and geotectonic relationships in northern Vermont and southern Quebec. Geol. Soc. Am. Bull., 71: 531–576.

Clark, T. H. (1952). Montreal area, Laval and Lachine map areas; Que. Dept. Mines, G. R. 46.

——— (1956). Oil and gas in the St. Lawrence Lowland of Quebec; CIMM Bull., Vol. 49 (531): 480–484.

Cloos, H. (1965). Results of explosion seismic studies in the Alps and in the German Federal Republic, in The Upper Mantle Symposium, New Delhi, edited by C. H. Smith and T. Sorgenfrei. Det Berlingske Bogtrykkeri Copenhagen, 94–103.

Cook, K. L. (1962). The problem of the mantle-crust mix: Lateral inhomogeneity in the uppermost part of the Earth's mantle; Advances in Geophys., 9: 296–371.

Diment, W. H. (1953). A regional gravity survey in Vermont, western Massachusetts and eastern New York; Ph.D. dissertation, Harvard Univ.

Dresser, J. A., and Denis, T. C. (1949). Geology of Quebec; Que. Dept. Mines, G. R. No. 20.

Ewing, G. N., Dainty, A. M., Blanchard, J. E., and Keen, M. J. (1966). Seismic studies on the eastern seaboard of Canada: The Appalachian system. Can. J. Earth Sci., 3: 89–109.

Fitzpatrick, M. M. (1953). Gravity in the Eastern Townships of Quebec. Ph.D. dissertation, Harvard University.

Garland, G. D. (1953). Gravity measurements in the Maritime Provinces. Pub. Dom. Obs. Canada, 16 (7).

——— (1955). Gravity measurements over the Cumberland Basin, N.S. Pub. Dom. Obs. Canada, 18 (1).

Gold, D. P. (1966). Alkaline ultrabasic rocks in the Montreal area, Quebec, in Ultramafic and related rocks, edited by P. J. Wyllie, Wiley, New York.

Goodacre, A. K. (1964a). A shipborne gravimetre testing range near Halifax, Nova Scotia. J. Geophys. Res., 69 (24): 5373–5383.

——— (1964b). Preliminary results of underwater gravity surveys in the Gulf of St. Lawrence. Dom. Obs. Canada, Gravity Map Series, No. 46.

Goodacre, A. K. and Nyland, E. (1966). Underwater gravity measurements in the Gulf of St. Lawrence, in Continental Drift, edited by G. D. Garland. Roy. Soc. Can. Spec. Publ., Toronto: University of Toronto Press: 114–128.

Houde, M. and Clark, T. H. (1961). Geological Map of St. Lawrence Lowlands. Quebec Dept. of Nat. Res., Map No. 1407.

Innes, M. J. S. (1960). Gravity and isostasy in northern Ontario and Manitoba; Pub. Dom. Obs., 21: (6).

Jones, I. W. (1962). Sedimentary basins and petroleum possibilities of Quebec. Proc. Geol. Assoc. Can., 14: 43–59.

Kay, M. (1942). Ottawa-Bonnechère graben and Lake Ontario homocline; Geol. Soc. Am. Bull., 53: 585–646.

Keen, C., and Loncarevic, B. D. (1966). Crustal structure on the eastern seaboard of Canada: studies on the continental margin. Can. J. Earth Sci., 3 (1): 65–76.

McConnell, R. K. (1966). Gravity and crustal rifting in the Canadian Shield; Pub. Dom. Obs.

MILLER, A. H. (1940). Investigations of gravitational and magnetometric methods of geophysical prospecting. Pub. Dom. Obs., *11* (6).

———— (1946). Gravimetric surveys of 1944 in New Brunswick; Geol. Surv. Bull., *6*: 1–28.

NETTLETON, L. L. (1941). Relation of gravity to structure in the northern Appalachian area; Geophys., *6*: 270–286.

SIMMONS, G. (1964). Gravity survey and geological interpretation, northern New York; Geol. Soc. Am., Bull. *75*: 81–98.

SOBCZACK, L. W. (1966). Gravity surveys in the Alexandria area, eastern Ontario; Pub. Dom. Obs. Canada (in press).

TANNER, J. G., and UFFEN, R. J. (1960). Gravity anomalies in the Gaspé Peninsula, Quebec. Pub. Dom. Obs. Canada, *21* (5).

THOMPSON, L. G. D., and GARLAND, G. D. (1957). Gravity measurements in Quebec, south of latitude 52°N. Pub. Dom. Obs. Canada, *19* (4).

WEAVER, D. F. (1966a). A geological interpretation of the Bouguer anomaly field of Newfoundland; Pub. Dom. Obs. Canada (in press).

———— (1966b). Preliminary results of the gravity survey in south-central Quebec and Labrador with maps; Dom. Obs. Canada, Grav. Map Series, Map Nos. 64, 65, 66, 67, 68.

———— (1966c). Preliminary results of the gravity survey of the island of Newfoundland; Dom. Obs. Canada, Grav. Map Series, Nos. 53, 54, 55, 56, 57.

WOOLLARD, G. P. (1962). The relation of gravity anomalies to surface elevation, crustal structure and geology; Res. Report Series, no. 62–9, USAF Aeronautical Chart and Information Center, St. Louis, Mo.

———— (1964). Bouguer gravity anomaly map of the United States; A. G. U. and U.S. Geol. Surv., Washington, D.C.

SOME GEOLOGICAL AND TECTONIC CONSIDERATIONS OF EASTERN CANADIAN EARTHQUAKES

W. E. T. Smith

ABSTRACT

About 1,500 earthquakes are known to have occurred in eastern Canada and adjacent areas. Their epicentres and magnitudes are currently being used as a basis for the preparation of maps from which seismic hazard may be estimated region by region. Isoseismal maps can be used to show the broad outlines of areas of differing stability but not to draw detailed geological conclusions. The Grand Banks earthquake afforded an excellent example of submarine transport of large masses of materials over great distances. No apparent correlation either in space or time could be found between earthquakes and faults.

SEISMOLOGISTS of the Observatories Branch, Department of Mines and Technical Surveys, have recently finished cataloguing all known Canadian earthquakes. One of the results of this project is a list of nearly 1,500 earthquakes of eastern Canada and adjacent areas lying east of longitude 85°W and between latitudes 40°N and 60°N. About half of these shocks were inside the borders of Canada. The list is published in two parts (Smith 1962, 1966). The earlier part contains the seismic history to the end of 1927. It is based on newspaper accounts, reports of religious, military, or civil officials, diaries and other private papers, scientific papers, and to a very small extent on long-period seismographs capable of recording only the larger tremors. The later part is an instrumental study of the locations and magnitudes of earthquakes from 1928 to 1959.[1] It is based on short-period sensitive seismographs, particularly those at Ottawa, Shawinigan Falls, Seven Falls, and Halifax, supplemented by data from stations at Cambridge, Burlington, Williamstown, Weston, and Fordham in the United States.

The precision to which epicentres[2] could be located varied with available data. In the historical work the position given is the central point of an area over which the shock was felt or, for the smaller ones, the municipality from which each was reported. The instrumental work is much more definitive. The majority of these epicentres are believed to be correct to within less than 20′ of arc. Little is known of the actual focal depths of the earthquakes. The relatively large radii over which they were felt led to speculation that they were rather deep; Lehmann (1955), however, investigated a few of the

[1]Subsequent data for the area were published by Milne and Smith (1961 to 1966).

[2]An epicentre is that point on the earth's surface directly above the focus of an earthquake.

larger ones and found that any conclusion of abnormal depth was unwarranted. A few focal depths within the crust have been given. These were estimated from travel time curves, but the procedure must be regarded as experimental.

Different methods were used in the two parts of the report to evaluate the sizes of earthquakes. In the first, the Modified Mercalli Intensity Scale was used. This is a scale of twelve degrees based upon perceptible effects, damage, etc. In the second the Instrumental Magnitude Scale of Gutenberg and Richter was applied. This scale is based upon measurements of the trace amplitudes recorded on calibrated seismographs. An empirical relation between these scales is given by the expression:

$$M = 1 + 2 \, I \, /3,$$

where M is the magnitude and I the maximum intensity. Though developed for average conditions in California, the formula applies reasonably well to northeastern North America. It is, however, a statistical relationship and some departure is to be expected in individual cases. A few comparisons are set out in Table I.

TABLE I

Earthquake	I (observed)	M (measured)	M (formula)
1925	IX	7.0	7.0
1935	VII	6.25	5.7
1944	VIII	5.9	6.3

The histograms in Figure 1 show the numbers of Canadian shocks of the various magnitudes, plotted in half-magnitude class intervals, for each of the two reports. The decrease in numbers with increasing size is normal, but the small numbers of very small shocks reflect inadequate coverage. The gap in the upper diagram occurred because conversion from integer intensities to magnitudes gives no values in this particular class interval.

In addition to the relationship between magnitude and intensity, Richter (1958) gives the following relation between magnitude and the seismic energy released:

$$\text{Log } E = 11.4 + 1.5M,$$

where E is the seismic energy in ergs. Benioff (1951) has reasoned that the numerical value of strain release in the material around the focus of an earthquake is proportional to the square root of the seismic energy released at the time of the earthquake. A knowledge of the magnitudes of earthquakes therefore permits the calculation of strain release. Figure 2 shows a plot of the cumulative sum of the square root of seismic energy known to have been released inside eastern Canada. The values are proportional to the cumulative sum of the strain released. The apparent increase in seismicity indicated by the trend is probably not significant since fluctuations are large because most of the energy is released by the larger earthquakes occurring

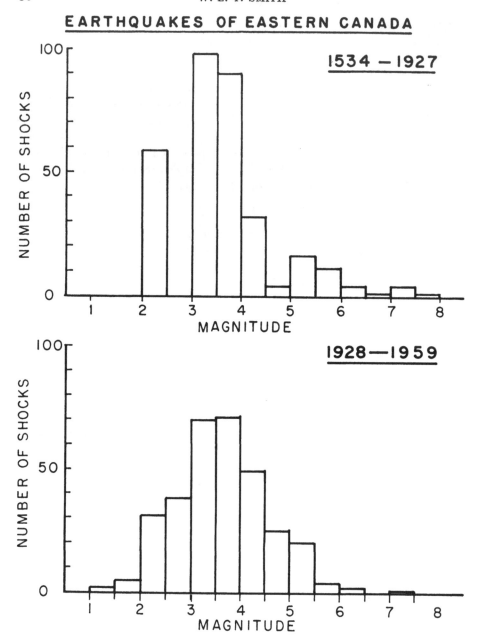

FIGURE 1. The numbers of Canadian shocks, of various magnitudes, plotted in half-magnitude class intervals.

at irregular intervals. Milne (1965) has used the data from the earthquake lists to map the strain release in the area as a step in the eventual production of detailed seismic regionalization maps from which earthquake hazard could be assessed for any part of the area. Seismic regionalization involves the combined use of all available geophysical and geological data and the

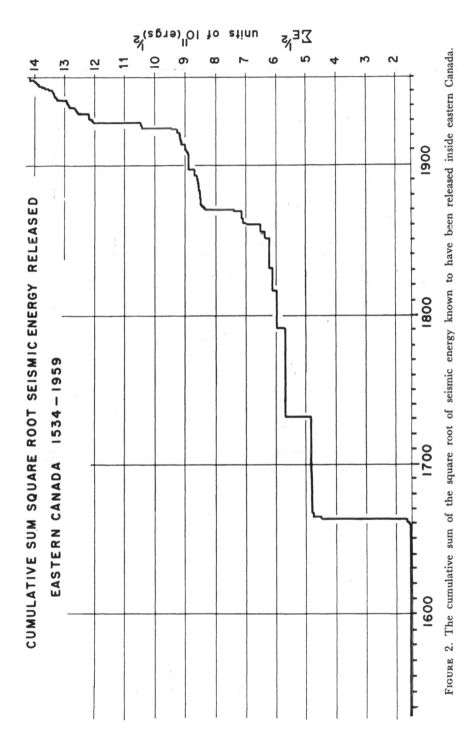

FIGURE 2. The cumulative sum of the square root of seismic energy known to have been released inside eastern Canada.

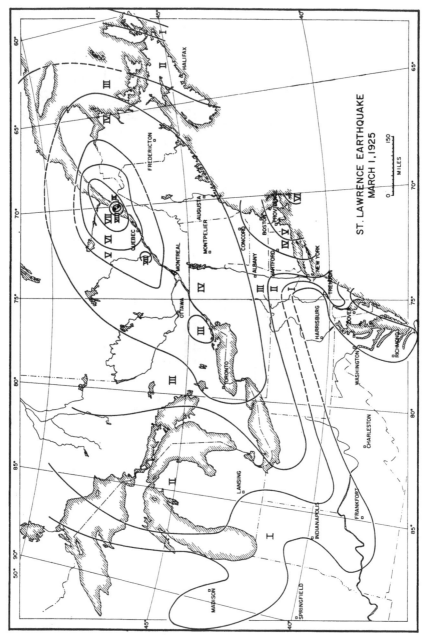

FIGURE 3. Isoseismal map of the St. Lawrence earthquake of 1925.

application of statistical methods to this data. The additional types of work necessary for this approach are carried on by other branches of the Department of Energy, Mines, and Resources. Consequently, a Departmental Committee on Seismic Regionalization has now been organized on a continuing basis.

The catalogues in their present form are finding increasing applications in the formulation of building codes, the design of large-scale power developments and actuarial studies for insurance rates. The purpose here is to examine them for evidence bearing on geology. Three obvious possibilities are: (1) that isoseismal[3] maps may delineate underlying structures through differences in the degree to which comparable areas are affected; (2) that recent disturbances may shed new light on geological processes; and (3) that earthquake epicentres may be correlated with known fault zones. Each will be considered in turn.

Figure 3 shows an isoseismal map of the St. Lawrence earthquake of 1925. This is one of six such maps in the second paper by Smith (1966). Isoseismals in United States territory are based on data collected by the United States Coast and Geodetic Survey and supplied to the Observatories Branch as part of a reciprocal arrangement. It can be seen that zones of low intensities—I, II, and III—occur relatively farther to the northeast in the Appalachian region while zones of higher intensities extend farther southwest along the coast. The effect in the Appalachians appears to be significant for, as Heck (1925) pointed out, there is an analogous bending of the isoseismal lines of certain United States earthquakes, which shows the same area to be one of relative stability. A further anomaly is the occurrence at Shawinigan Falls of intensity VII surrounded by intensity V. Here many stone and brick walls, though well built, were cracked because the buildings were placed on or near the slopes of clay banks. The intervening isoseismal was not defined by the available data. Notwithstanding these observations, it must be admitted that, in general, isoseismal maps of past Canadian earthquakes are not suitable to support detailed geological conclusions. The distribution of populated places tends to be along rivers and railways, with few settlements elsewhere and large uninhabited areas to the north. It follows that data (even if plentiful) from replies to special questionnaires are in the nature of a series of "traverses" (not necessarily in optimum positions) across the area of interest. Under such conditions detail will be missed in resulting maps.

The Grand Banks earthquake of 1929 furnished unusual evidence of the transport of large masses of material over great distances along the ocean bottom. Twelve transatlantic cables were broken, each in at least two places, for a total of twenty-eight known breaks. Their positions are shown in Figure 4, adapted from Doxsee (1948). For 17 hours and 13 minutes following the shock, there occurred an orderly sequence of breaks, each succeeding broken cable lying in increasingly deeper water extending

[3]Isoseismals are lines separating areas of different intensity.

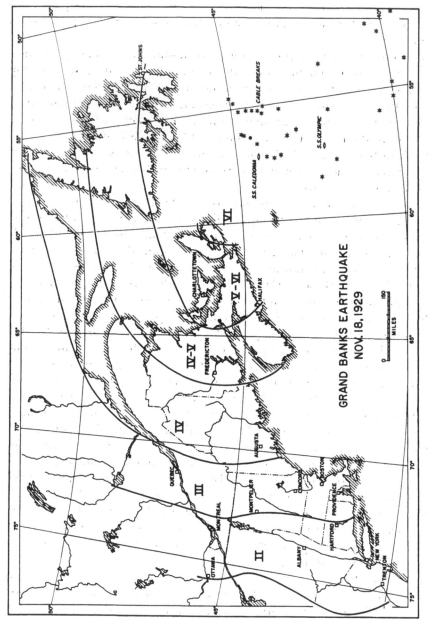

FIGURE 4. Isoseismal map of the Grand Banks earthquake of 1929.

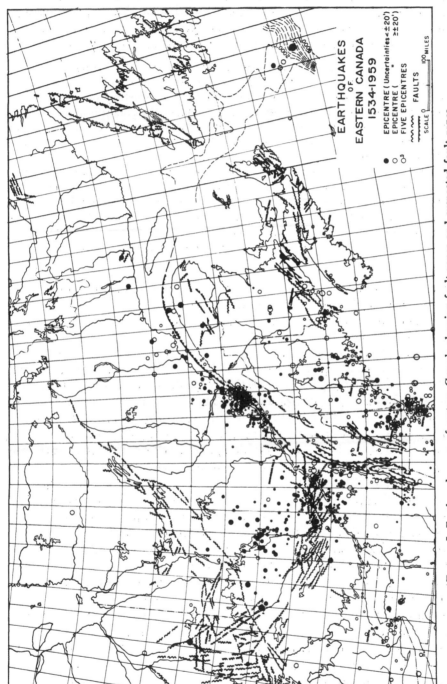

FIGURE 5. Earthquake map of eastern Canada showing fault zones and suspected fault zones.

more than 300 miles south of the epicentral area. Many attempts were made to explain the cause of the cable breaks, usually in terms of active faulting. None of these accounted satisfactorily for the orderly sequence of the breaks until Heezen and Ewing (1952) proposed the following:

[The] severe shock jarred the continental slope and shelf, setting landslides and slumps in motion. . . . The mass movements, starting on the relatively steep continental slope, raced downwards, and by the incorporation of water the moving sediment was transformed from sliding masses into turbidity currents. . . . The currents had many times the force necessary to break cables, and snapped each cable shortly after reaching it. In all cases at least two breaks occurred 100 miles or more apart, and the intervening section of cable was either buried or carried seaward so far that it was never found. As the current began to slow down because of decreasing slope, finer and finer sediments were deposited in a manner which produced graded bedding. Since the flow passed the last cable in the area at a distance of about 400 miles from its origin with a velocity of 12 knots and sufficient force to destroy 200 miles of cable it is probable that the area of deposition extended hundreds of miles farther.

The epicentre of this earthquake was recently recomputed through the courtesy of the United States Coast and Geodetic Survey using the facilities of their epicentre-determination programme and data from 119 seismograph stations. It is of interest in connection with the above thesis that the new position was found to be at 44°.5N, 56°.3W, where the channel of Cabot Strait reaches the steep slope of the continental shelf and the water is 4,000 to 5,000 feet deep (see Fig. 5). This is about 80 miles southwest of the earlier position found by Doxsee (1948).

Figure 5 is an earthquake map of eastern Canada upon which have been sketched the fault zones and suspected fault zones of the area from the Tectonic Map of Canada, 1950, prepared by the Geological Association of Canada with the support of the Geological Society of America. The submarine contours are from the same source and a few additional faults are from tectonic maps prepared by the Geological Survey of Canada. It may be noted that there is a large concentration of earthquakes along the St. Lawrence river below Quebec City and another broad band extending down the Ottawa and Gatineau rivers through the Ottawa-Montreal area and across the United States border. As the seismographs at Ottawa, Shawinigan Falls, and Seven Falls contributed to the location of epicentres in both these regions, the large stable area between them is a highly significant feature of the map. This is also true of the relative stability of the Maritime provinces, for this area was settled early. The lack of earthquakes to the north may well be because the area was neither settled nor monitored by seismograph stations. It can, however, be said that no very large earthquake has occurred there since the turn of the century when seismographs came into use. A seismograph station has only recently been installed in Newfoundland (St. John's) and a historical study of the earthquakes of this province has yet to be made.

There is no obvious correlation between the positions of the faults and the earthquake epicentres. There are many faults where there are no known earthquakes and many earthquakes where there are no known faults. Places where faults and earthquakes do coincide (to within the limits of their precision) are apparently not more numerous than might be expected by chance. This is, perhaps, not very surprising for the ages of the faults are reckoned in millions of years, while the known earthquakes date back only a few hundred. This lack of correlation limits the usefulness of the known faults in seismic regionalization mentioned earlier. The faults will still have to be considered in connection with foundation conditions but are of no use in estimating where earthquakes in eastern Canada are likely to occur.

While, admittedly, little is known of the depths of focus of these earthquakes except that they are not abnormally deep (much below 25 km), it can also be said that none has caused surface faulting. The fractures, if any, must therefore lie at some intermediate depth. J. H. Hodgson of the Observatories Branch has attempted fault-plane solutions for some of the larger shocks to try to determine the direction of faulting at their foci. He was unsuccessful because of the low sensitivity of the seismographs on which they were recorded. Despite the lack of obvious correlation between faults and earthquakes, the possibility cannot be entirely ruled out that the earthquakes below Quebec City are connected with changes in structure at depth across the Appalachian front. The Geodetic Survey reported to the Seismic Regionalization Committee that level lines, recently rerun after 30 years, showed settling of the land in the Lake St. John area relative to Quebec City. So far, no relation has been established between this slow vertical movement and the seismicity of the area. No evidence of recent horizontal movement was reported.

REFERENCES

BENIOFF, H. (1951). Earthquakes and rock creep. I, Seismol. Soc. Am. Bull., *41* (1): 31–62.

DOXSEE, W. W. (1948). The Grand Banks earthquake of November 18, 1929, Dom. Obs. Pub., *7* (7): 323–335.

HECK, N. H. (1925). Earthquakes of 1925, Seismol. Soc. Am. Bull., 15(2): 106–113.

HEEZEN, B. C., and EWING, M. (1952). Turbidity currents and submarine slumps, and the 1929 Grand Banks earthquake, Am. J. Sci., *250*: 849–873.

LEHMANN, I. (1955). The times of P and S in northeastern America, Annali di Geofisico, *8* (4): 351–371.

MILNE, W. G. (1965). Earthquake risk in Canada, Ph.D. thesis, University of Western Ontario: 1–233.

MILNE, W. G., and SMITH, W. E. T. (1961). Canadian earthquakes—1960, Seismol. Ser. Dom. Obs., 1960–2: 1–23.

——— (1962). Canadian earthquakes—1961, Seismol. Ser. Dom. Obs., 1961–4: 1–24.

——— (1963). Canadian earthquakes—1962, Seismol. Ser. Dom. Obs., 1962–2: 1–22.

——— (1966). Canadian earthquakes—1963, Seismol. Ser. Dom. Obs., 1963–4: 1–30.

RICHTER, C. F. (1958). Elementary seismology, W. H. Freeman and Company, Inc., San Francisco.

SMITH, W. E. T. (1962). Earthquakes of eastern Canada and adjacent areas, 1534–1927, Dom. Obs. Pub., *26* (5): 269–301.

SMITH, W. E. T. (1966). Earthquakes of eastern Canada and adjacent areas, 1928–1959, Dom. Obs. Pub., *32* (3): 87–121.

SOME IMPLICATIONS OF NEW IDEAS ON OCEAN-FLOOR SPREADING UPON THE GEOLOGY OF THE APPALACHIANS

J. T. Wilson, F.R.S.C.

ABSTRACT

It has long been known that the sun's magnetic field dies away about every eleven years and returns with reversed polarity. For some time geomagneticians have suspected that the earth's field behaves in the same way and it is now well established that there have been ten reversals during the past three and a half million years.

Recently Vine and Matthews pointed out that this could explain the regular patterns of anomalies parallel to mid-ocean ridges if the ocean floors were expanding and being imprinted magnetically by the reversals.

This can provide a precise time-scale for continental drift and makes its reality far more plausible. The rate of drift indicated could hardly be accounted for unless as some oceans spread others diminished, presumably by absorption into trenches and beneath island arcs.

Kay has suggested that a series of arcs lay off the coast of eastern North America in Lower Paleozoic time and the writer suggests that a proto-Atlantic Ocean closed along these arcs merging "Atlantic" and "Pacific" faunal realms of which alternating examples are scattered along both coasts of the present ocean. This paper points out additional evidence in support of this hypothesis, especially the discovery of European-type trilobites in Florida by H. B. Whittington, and that there is a difference in age of the basement beneath Cambrian rocks with fossils of "Atlantic" and "Pacific" faunal realms.

ANY REGIONAL STUDY such as this one of the Appalachians immediately comes face to face with the basic problem of the science of the solid earth which is that a vast amount of excellent detailed data is available, but that no good theory is agreed upon and this is necessary if the data are to be summarized.

The objective of most geologists and geophysicists is to make many observations over a small area and either produce a map or find an economic deposit or both. To discuss a larger area briefly, something quite different is required. What is needed is a synthesis of the geological history, not a library full of the original observations. In order to make this summary a theory of some sort is indispensable. Now, geological theories have been very vague affairs which are wisely eschewed by most as "armchair geology." Geophysical theories, such as those of Sir Harold Jeffreys, are precise, but they apply to models which are so grossly simplified that they bear little relation to the real earth.

At the heart of the matter has been the inability to decide upon the extent of past movement in the earth. Although folding shows that some motion has occurred, the natural view has been to minimize this and, in particular, to hold that the continents remained fixed. This assumption has not been satisfactory, because it has not served to explain geology well. At best our ideas about such matters as mountain building are still vague. If our basic assumption that the earth as a whole has been fairly rigid is wrong, our ideas need revision and much of what has been written about historical and structural geology, tectonics, petrogenesis and physics of the earth is out of date.

Fifty years ago, A. Wegener (1) pointed out three reasons for believing that parts of the crust are being displaced relative to other parts by movement over some mobile layer. These points were the excellent fit of some opposite coasts, the rise of land after ice sheets melted and the displacement of climatic zones of the past. Most geologists and geophysicists did not accept these views, but Wegener's arguments were never wholly dismissed. Several recent discoveries bear on the problem. The earth's magnetic field is reversing every few hundred thousand years (2, 3). The paleomagnetism locked in rocks at the time of their formation suggests that large displacements of the crust have taken place (4). As M. Ewing and H. W. Menard and their colleagues have done so much to show, the mid-ocean ridge is the greatest mountain system on earth, quite different from continental mountains and not explained by existing theories of a rigid earth (5). The crust may be growing and expanding away from it (6, 7). The geometry of fractures formed between moving crustal plates demands new kinds of faults, named transform faults, with quite different properties from normal, transcurrent, or thrust faults (8). These discoveries taken together have in the past few months opened the possibility of resolving the dilemma and give prospect of transforming the accounts of the later part of geological history from vague generalities to a precise quantitative basis.

The present advance stems from the suggestion by F. J. Vine and D. H. Matthews (9) at Cambridge and, independently, by L. W. Morley and A. Larochelle (10) that if the ocean floors are spreading and if the earth's magnetic field is reversing then every reversal should be discernible in magnetic anomalies over oceans. F. J. Vine and J. T. Wilson (11) discovered off California the first example of a ridge which appears to be expanding and used it to illustrate the theory. They interpreted the anomalies to deduce the age, total displacement, and rate of motion along the San Andreas rift, which is a transform fault. F. J. Vine (12) and J. R. Heirtzler (13) have shown by other examples that magnetic anomaly patterns of the ocean basins can be precisely explained if reversals of the earth's field influence the magnetization of lavas poured out along expanding mid-ocean ridges. The rates of expansion have been quite steady for the past four million years, being 1.9 cm/yr south of Iceland, 6 cm/yr off British Columbia, and 9 cm/yr in the South Pacific. This interpretation has been accepted by E. A. Godby et al. (14), while J. R. Heirtzler at Lamont Geological Observatory is attempting to extend it to all oceans.

This method promises to give a precise picture of where every continent lay at the time of every reversal (about every million years) back to Jurassic time. If this is true the later history of continental drift changes from being a vague and dubious hypothesis to being a numerically precise design for recent geology.

Doubts that the mid-ocean ridge is expanding are further removed by an abstract now in press (15) for the next meeting of the Geological Society of America. L. Sykes (16) of Lamont has measured the direction of motion of earthquakes along offsets in the mid-ocean ridge. Theory shows that they should be in opposite directions for transcurrent and transform faults. The mechanisms of about twenty earthquakes on mid-ocean ridges have been studied and he concludes that "the sense of strike-slip motion is in agreement with that predicted for transform faults and for various hypotheses of ocean-floor growth; it is opposite to that expected for a simple offset of the ridge crest along the various fracture zones."

That drift was going on before the Mesozoic era seems probable and J. T. Wilson (17, 18) has recently applied this motion to suggest that an early Atlantic Ocean closed during the Paleozoic and has reopened again since the Jurassic. The mechanism which would give rise to these great shifts seems likely to be convection in a shallow layer as advocated by W. M. Elsasser (19), D. C. Tozer, and E. Orowan (20) in recent papers. Preprints just received of a series of papers by A. E. Ringwood's group at Canberra (20) discuss the petrological aspects of such a system including the origin of basalts, andesites, and anorthosites.

It is only very recently that my ideas on the application of this cyclical drift to the Appalachians were published. Under the circumstances I can only add a few notes to those papers (17, 18) in which I suggested that an open Atlantic existed in Cambrian and Lower and Middle Ordovician time which completely separated the "Atlantic" and "Pacific" faunal realms. Kay's island arcs lay along the whole western and southern coast of this ocean which slowly closed when the arcs overrode the ocean floor along the trenches in front of these arcs.

During the Middle Ordovician the first meeting of the two coasts led to an exchange of fauna and the end of the marked difference between the two realms. The coasts had different shapes and, as promontories met and overlapped, mountains were raised, producing subaerial fans of red-beds on either continent. The ocean gradually disappeared and each continent became the borderland of the other in the sense that Schuchert and Barrell have discussed. When the ocean had closed completely the arid climates of Permian, Triassic, and Jurassic time followed. Not until the end of the Jurassic did the Atlantic reopen in the latitudes of the United States coast and not until early Tertiary time did it open in the north. When it did so, some coastal fragments were transposed and these include the coast of Norway, Scotland, and Northern Ireland which have "Pacific" faunas in their Lower Paleozoic rocks. Part of what had been Europe became joined to North America to form a coastal strip with "Atlantic" faunas from eastern Newfoundland to Connecticut.

Sougy had already drawn attention to what may be the next fragment transposed from North America to Africa, while the presence of flat-lying Paleozoic sedimentary rocks beneath Cretaceous cover in central Florida led me to suggest that this part of Florida had once been attached to Africa. It must have formed on the southeast side of the Paleozoic island arcs and hence belonged on the other side of the Paleozoic ocean from most of North America. Through the help of Carol Faul, H. G. Richards, and H. B. Whittington I can now support this view with the description by Whittington (22, 23) of a trilobite found in one drill-hole in Florida.

Of special interest is its relationship to trilobites of central and southern Europe and northwest Africa (not with any so far known in North America). . . . The Florida trilobite seems to be a lone representative of this more southerly fauna, and its presence may indicate that the Florida Palaeozoic rocks were laid down in a province faunally separated from that of the Appalachian-Ouachita trough. In addition Prof. H. J. Harrington, University of Buenos Aires, informs me that the Florida trilobite is unlike any known in South America (23).

I have also examined the aeromagnetic maps published by the Geological Survey of Canada for southern Newfoundland. Although the general trend of the anomalies throughout the region strikes southwest, there is an especially well-marked change in the anomaly pattern along the line of separation of the continents through Freshwater and Hermitage bays which the geological work of F. D. Anderson (24) and H. Williams (25) had already suggested to be the junction of different provinces.

An interesting point which has recently occurred to me is that the ages of basic intrusives along the North Atlantic coasts show two marked peaks. In Scotland, Ireland, east and west Greenland, and Baffin Island the basic lavas and intrusives range in age from 50 to 60 m.y., whereas from Nova Scotia through New York and North Carolina to Florida the age of the late volcanic rocks along the coast is from 140 to 180 m.y. An intrusive near Lisbon is 85 m.y. Thus if these rocks date the onset of ocean rifting, it would seem that the ocean had opened in two stages with the line of separation probably passing close to Newfoundland and Gibraltar. This possibility should be considered in interpreting the geology of Newfoundland and the Grand Banks.

Phillip B. King has prepared a map of basement provinces (private communication). He has distinguished a very late Precambrian province (800 to 500 m.y.) from the somewhat older Grenville Province. The younger province extends through eastern Newfoundland, Nova Scotia, the coast of New Brunswick, and eastern New England precisely beneath those regions which have a Cambrian-Ordovician fauna having affinities to Europe. It is interesting that whereas the ages of the Precambrian of Greenland and

Scotland fit those of Labrador, those reported from inliers in England are from 485 to 600 m.y. and fit those of the North American east coast. The Grenville is not found in the British Isles (5).

The situation is thus that an account has now been given of the closing and opening of the North Atlantic region, and the fit of Africa to South America is well known. Great uncertainty still remains about the history of the Caribbean and Tethys regions and about the possible junctions of continental fragments along the Amazon Valley and across North Africa.

Perhaps some of these areas closed later and the disturbance is marked by the later (post-Devonian) shear faults through the Great Glen, Newfoundland-Maritime-New England region (Cabot fault), and the Southern Appalachians (Brevard fault zone).

The problems raised are so numerous and important that no discussion of the Appalachian region can afford to continue to ignore the probability of drift.

REFERENCES

1. WEGENER, A. Origin of continents and oceans, Dutton, New York (1924).
2. COX, A., DOELL, R. R., and DALRYMPLE, G. B. Reversals of the earth's magnetic field, Science 144 (1964): 1537.
3. MACDOUGAL, I. and TARLING, D. Dating geomagnetic polarity zones, Nature 202 (1964): 171.
4. IRVING, E. Paleomagnetism, New York: Wiley (1964).
5. BLACKETT, P. M. S., BALLARD, E. C., and RUNCORN, S. K. (eds.). Symposium on continental drift (Esp. papers by B. C. Heezen and H. W. Menard), Phil. Trans. Roy. Soc., London, A258 (1965).
6. HOLMES, A. Radioactivity and earth movement, Trans. Geol. Soc., Glasgow, 18 (1928–29): 559.
7. HESS, H. H. Mid-ocean ridges and tectonics of the sea floor in W. F. Whittard and R. Bradshaw, eds., Submarine Geology and Geophysics, London: Butterworth (1965).
8. WILSON, J. T. A new class of faults and their bearing on continental drift, Nature 207 (1965): 343.
9. VINE, F. J. and MATTHEWS, D. H. Magnetic anomalies over oceanic ridges, Nature 199 (1963): 947.
10. MORLEY, L. W. and LAROCHELLE, A. Paleomagnetism as a means of dating geological events, in Geochronology in Canada, edited by F. Fitz Osborne, Roy. Soc. Can., Spec. Publ. 8 (1964): 39.
11. VINE, F. J. and WILSON, J. T. Magnetic anomalies over a young oceanic ridge off Vancouver Island, Science, 150 (1965): 485.
12. VINE, F. J. Spreading of the ocean floor; new evidence. Science, 154 (1966): 1405.
13. PITMAN, III, W. C. and HEIRTZLER, J. R. Magnetic anomalies over the Pacific-Antarctic Ridge, Science 154 (1966): 1164.
14. GODBY, E. A., BAKER, R. C., BOWER, M. F., and HOOD, P. J. Aeromagnetic reconnaissance of the Labrador Sea, J. Geophys. Res., 71 (1966): 511.
15. WILSON, J. T. Did a lower Palaeozoic ocean through New England close and then reopen to form the Atlantic? Geol. Soc. Amer., Northeastern Section. Abstracts (1967): 66, 67.
16. SYKES, L. Mechanism of earthquakes and nature of faulting on the mid-oceanic ridges, Geol. Soc. Am. Abstracts, 1966 Annual Meeting: 216, 217.
17. WILSON, J. T. Did the atlantic ocean close and reopen? Nature, 211 (1966): 676.
18. WILSON, J. T. Are the structures of the Caribbean and Scotia Arc regions analogous to ice rafting? Earth and Planet. Sci. Letters 1 (1966): 335.

19. ELSASSER, W. M. Thermal structure of the upper mantle and convection, *in* Advances in earth science, *edited by* P. M. Hurley, M.I.T. Press, 1966.
20. OROWAN, E. Convection in a non-Newtonian mantle, continental drift, and mountain building, Phil. Trans. Roy Soc. London, *A258* (1965): 286.
21. RINGWOOD, A. E. *et al.*, Petrology of the upper mantle (7 preprints bound together), Australian Nat. Univ., Publ. *444* (August 1966).
22. WHITTINGTON, H. B. A new Ordovician trilobite from Florida, Breviora, *17* (1953): 1.
23. WHITTINGTON, H. B. Phylogeny and distribution of Ordovician trilobites. J. Paleont. *40* (1966): 696.
24. ANDERSON, F. D. Belloram sheet, Geol. Surv. Can., Map 8–1965 (1965).
25. WILLIAMS, H. The Appalachians in northeastern Newfoundland: a two-sided symmetrical system, Amer. J. Sci. *262* (1964): 1137.

Lightning Source UK Ltd.
Milton Keynes UK
UKHW030613210722
406167UK00006B/671